Vulnerability and Risk Measurement of Climate Induced Disasters in Gujarat

Published by
ACADEMIC FOUNDATION
in association with

Sardar Patel Institute of Economic & Social Research,
Ahmedabad

About the Author

Shital Shukla is associate Professor (Geography), M.S. University of Baroda, Vadodara. She is also a faculty (on Lien) of Sardar Patel Institute of Economic and Social Research, Ahmedabad. She is PhD from Gujarat University and her doctoral thesis was entitled: *Sustainable Development of the Coastal Region of Gujarat*. Her areas of research interest are: natural resource management, climate change and disaster management, regional planning, human development, sustainable development and environmental planning.

She is deeply involved in projects related to human development, sustainability and livelihood issues, climate change and disaster risk management and urban water disasters. She has presented many research papers in national and international seminars and conferences. She has worked with UNDP, DFID, International Cooperative Alliance, National Disaster Management Agency, National Institute of Disaster Management and Ministry of Human Development

Vulnerability and Risk Measurement of Climate Induced Disasters in Gujarat

SHITAL SHUKLA

ACADEMIC FOUNDATION
NEW DELHI

www.academicfoundation.com

First published in 2014
by

ACADEMIC FOUNDATION
4772-73 / 23 Bharat Ram Road, (23 Ansari Road),
Darya Ganj, New Delhi - 110 002 (India).
Phones : 23245001 / 02 / 03 / 04.
Fax : +91-11-23245005.
E-mail : books@academicfoundation.com
www.academicfoundation.com

In association with

Sardar Patel Institute of Economic & Social Research, Ahmedabad

Disclaimer:
The findings/views/opinions expressed in this book are solely those of the author and do not necessarily reflect the views of the publisher.

Cataloging in Publication Data—DK
Courtesy: D.K. Agencies (P) Ltd. <docinfo@dkagencies.com>

Shukla, Shital.
Vulnerability and risk measurement of climate induced disasters in Gujarat / Shital Shukla.
p. cm.
Includes bibliographical references (p.).
ISBN 9789332701069

1. Climatic changes—Environmental aspects—India—Gujarat. 2. Natural disasters—India—Gujarat. 3. Emergency management—India—Gujarat. 4. Risk assessment—India—Gujarat. I. Title.

DDC 551.695475 23

Typeset by Italics India, New Delhi.

Printed and bound by The Book Mint, New Delhi.
www.thebookmint.in

Dedicated to my parents

Mr V.H. Lodhia (Late) and Mrs T.V. Lodhia
who have inspired and motivated me

Contents

List of Tables, Figures and Annexures

Tables

Figures

Annexures

Abbreviations

AIDMI	All India Disaster Mitigation Institute
BBC	Bogardi, Birkmann and Cardona
BMTPC	Building Materials and Technology Promotion Council
BPL	Below Poverty Line
CMIE	Centre for Monitoring Indian Economy
CRED	Centre for Research on the Epidemiology of Disasters
CRF	Calamity Relief Fund
CRS	Calamity Relief Services
DDP	District Development Programme
DPAP	Drought Prone Area Development Programme
EM-DAT	International Database Series
FWWB	Friends of Women's World Banking
GDI	Gender Development Index
GDP	Gross Domestic Product
GEI	Gender Empowerment Index
GoG	Government of Gujarat
GoI	Government of India
GIDM	Gujarat Institute of Disaster Management
GR	Growth Rate
GSDMA	Gujarat State Disaster Management Authority
GSDP	Gross State Domestic Product
HDI	Human Development Index
HH	Household

IDRS	Integrated Disaster Reduction Strategies
IMD	Indian Meteorological Department
IMR	Infant Mortality Rates
IPCC	Intergovernmental Panel on Climate Change
ML	Medium and Large
MNREGA	Mahatma Gandhi National Rural Employment Guarantee Act
NABARD	National Bank for Agriculture and Rural Development
NCCF	National Calamity Contingency Fund
NDRF	National Disaster Relief Fund
NFCR	National Fund for Calamity Relief
NGOs	Non-Governmental Organisations
NSDP	Net State Domestic Product
OFDA	Office of U.S. Foreign Disaster Assistance
PRA	Participatory Rural Appraisal
RR	Revenue Receipt
SC/ST	Scheduled Castes and Scheduled Tribes
SD	Standard Deviation
SDMA	State Disaster Management Authority
SDRF	State Disaster Relief Fund
SEWA	Self Employed Women Association
SPSS	Statistical Package for Social Science
SRS	Sample Registration System
SS	Small Scale
TDS	Total Dissolved Salts
TFR	Total Fertility Rate
UNDP	United Nations Development Programme
UN/ISDR	United Nations International Strategy for Disaster Reduction
US	United States

Foreword

This is an interesting and timely book. It is on disasters and vulnerability and is rooted in Gujarat. It discusses the literature, gives a lot of background material on Gujarat, and then gets into disasters and vulnerability. It gives details on phenomenon like floods, cyclones and so on. Periodicity, losses and other details are also spelt out. There is a major focus on policies. I won't repeat the arguments; for the reader must go for it her/himself. Instead I will add on by discussing vulnerability and poverty as concepts, the agro-climatic structure of the Gujarat economy in terms of soil and climate, and the emerging volatility on account of climate change for Gujarat's coastline.[1] We do this by outlining the delicate combination of the process of siltation and sea currents leading virtually to the establishment of sea ports in Gujarat and physical and hydrological changes in them. This kind of balance is the swan song of this book and we just give one more example. While geological data could have been used, we give historical descriptions of physical changes with flooding of rivers particularly from Lothal. We then describe the changes taking place in river flows in Gujarat from modelling studies and show that the process of siltation and the sustainable balances will be seriously undermined as river flows decline. Climate change and sea warming will lead to rise in sea levels. These processes were earlier expected over longer periods but reports presented at the Doha climate change conference suggest a significant hastening of the process. Gujarat has developed its ports very well in the last decade but these trends could seriously jeopardise the process. Policy action will have to be at much higher levels than in Gujarat or for that matter even the country.

1. This literature was presented in a seminar on Gujarat's port towns arranged by Shri Hasmukh Shah, former Chairman of the Gujarat Ecology Commission.

Vulnerability and Poverty

It is mainly the poor that suffer from famine, hunger and malnutrition. But not all poor people are equally vulnerable to hunger, and it is not always the poorest that are exposed to the greatest risks. There are many other factors that determine the vulnerability to hunger. These include the risk of exposure to crises, stress and shocks. There is also the population which is subjected to risk by civil breakdowns and violence on a community scale.

As the Chairman of the Advisory Committee of the Gujarat Disaster Management Authority set up after the Kutch earthquake, I argued that the fast development the State has gone through, without corresponding development in civil society organisations can lead to greater vulnerability. If Gujarat's manufacturing sector doubles every seven years as it has in the past and development takes place on a highly decentralised scale as it has, with every district having more than ten thousand workers in the manufacturing sector, when disaster like an earthquake or civil violence hits, the damage is far more. Earthquakes are natural but damage depends on land use, the nature of development and on how social and governance institutions cope up.

As vulnerability is not coincident with poverty, malnutrition or other indices of human deprivation, more research should be based on a better understanding of such environmental crises as this book argues. Also, public responses to improve the coping facilities of social groups should be assisted by targeted research and development projects, and not just by relief. A number of approaches have been chosen to address these problems (e.g., entitlement, empowerment, enfranchisement). Food security should be closely tied to human security. While food supply to the rural poor is still a major problem, more emphasis on urban food security is needed.

The general tendency to combat environmental and social forms of degradation as expressed in famine and poverty should not only be "reactive" but should stress on preventive measures. The term "preparedness" should move into the centre of attention. This leads to the increased necessity of early warning systems. Policymakers have already reacted by establishing short and medium term early warning systems for the most vulnerable regions. The Famine Early Warning System being developed by the United Nations is an example of such a reaction. However, without accurate socioeconomic and political information from the field, such computer assisted warning systems based mainly on biophysical parameters, will not be able to accurately predict famines and social breakdown in a given region.

We now turn to another problem, an ecologically fragile economic sector in this case enmeshed in climate change.

Ports and Climate

Given the physical and bio-genetic diversity of Gujarat, it makes it economically rewarding and resource augmenting, to follow a strategy of diversified and regionally differentiated development. For example, in the agro-ecological zoning of the FAO, Gujarat (see Table 1) and also India is (are) one of the few developing countries (areas) studied which has development potential in almost all climate types—dry, semi–moist, semi-arid, sub-humid, fluvisols and gleysols. Within it are seen the agro-ecological zones of the developing world. This is indeed a great policy challenge and

Table 1

Agro-Climatic Regions of Gujarat

Region	*District*	*Rainfall (in mm)*	*Climate*	*Soils*
1. Southern Hills	Dangs, Bulsar	1793	Semi-arid dry	Black coastal sub-humid deep alluvium
2. Southern Gujarat	Surat, Bharuch	974	-do-	Black coastal alluvium
3. Middle Gujarat	Baroda, Kheda Panchmahals	904	Semi-arid	Medium black
4. North Gujarat	Ahmedabad, Gandhinagar, Mehsana, Sabarkantha, Banaskantha	735	Arid to semi-arid	Gray brown coastal alluvium
5. North West	Kutch	340	Arid	Gray brown deltaic alluvium
6. North Saurashtra	Amreli, Bhavnagar, Jamnagar, Rajkot, Surendrangar	537	Semi-arid	Medium black calcareous
7. South Saurashtra	Junagadh	844	Dry Sub-humid	Coastal alluvium, Medium black

Source: Alagh (1989).

opportunity. This is particularly so in an emerging environment which regards bio-diversity as nature's bounty and not only a constraint to technological progress. Climate change should also be seen in this context. Nowhere is it truer than in Gujarat. Unfortunately the present agro-climatic zones of Gujarat do not separately treat the coastal zones.

Through the millennia, siltation from the great rivers which flow through Gujarat, the Sabarmati, the Mahi, the Narmada and the Tapi, have brought enriching silt and virtually created the soils which have led to reclamation of land from the sea. This is also true of the impact of the hydrological flows of the Saurashtra Plateaux. There are many detailed studies and descriptions of this phenomenon. Given the historical backdrop of the meeting, we used the classic volumes on the excavations at Lothal by the Archaeological Survey of India (ASI). However, equally vivid use can be made of the Cambridge Baroda Archaeological Project at Champaner, which give evidence of the receding of the sea. The ASI two volume magnum opus on Lothal (ASI, 1979; 1985) is full of discussions of hydrological flows and river floods and silt.

The Gujarat Ecology Commission's (2012) popularising volumes on the ports of Gujarat also suggest the impact of the reclamation of land from the sea having an impact on port town development and flora and fauna. We do not seem to have easily available studies on the macro effects of the net accreting effects on delta sedimentation of Gujarat's rivers but for Bangladesh, Broadhaus (2001: 355) has made estimates of the 'roughly stable or in net accreting' land area in the Ganga and Brahmaputra deltas.

There is systematic evidence to show that river flows are declining in Gujarat deltas and will continue to do so (Gupta *et al.*, 2009). The following estimates show this trend (Table 2).

Table 2

Reported and Estimated Runoff

No.	*River Basin*	*Reported Runoff 1975 Mha-m*	*Estimated Normal 1951/80 Mha-m*	*Projected 2080; A2 Mha-m*
1	Sabarmati	0.4	1.4	0.39
2	Mahi	1.2	2.0	0.83
3	Narmada	4.0	5.0	5.14
4	Tapi	2.0	2.7	1.18

Source: Gupta *et al.*, 2009: 172

This phenomenon can be linked to the shortage of water that is emerging in India. A study for 2020 shows that a high growth scenario is not feasible in a BAU sense because water becomes a binding constraint (see Alagh, 2006), Water conserving scenarios lead to substantial income generation outcomes for the agricultural sector in another study (Alagh, 2011).

Global Warming and Erosion

The global warming scenarios will have a reverse impact on accretion of land and soil, and this impact will be adverse. At present, this is dimly understood in terms of the future of port towns, their populations, cultural and economic development. Gujarat's ports have grown very fast in the last two decades. Some of them like Dahej are environmental hot spots. Global warming may impact on these ports is not talked of although, as we all know, the Intergovernmental Panel on Climate Change, 2007 highlighted 'coasts' as one of the five key impacted areas of global warming. There are a number of studies on sea level rise on account of warming. S. Suresh Kumar (2012: 259) in a recent review concludes, "In India the studies have revealed that sea level rise is by 1.06 to 1.75 mm per annum." More recently for the Doha Conference, The Potsdam Institute's report, according to newspaper reports, suggests much higher sea level rise, but these studies have yet not received critical assessments. Regional impacts could be higher (Ray *et al.*, 2009), for example measured climate change in Gujarat for the last three decades and show impacts on temperature and rainfall. IIT Kharagpur's P.K. Nath (2012: 82) amongst others has looked at impacts of sea warming on coastal communities and in his words, "Erosion, submergence of shorelines, salinity of the water table due to increased sea levels would have serious effects." The Gujarat Ecology Commission's (2012) rapid surveys of the coastline of Gujarat also shows sea erosion as a challenge in Dandi, Ghogha, Gopnath, Mahuva, Porbandar, Ubharat, Umbergaon.

There is evidence to suggest that climate change will have definite impacts on port towns in Gujarat and the communities who live in them and live off them. Some analysts build horror scenarios of migration of vast populations. A minimalist approach is to study the phenomenon. It is interesting that Gujarat institutions are not represented say in the Green House Emissions Assessments for India (INCAA, 2007; 2010). It is obvious that the State should play a proactive role in the serious work the country is doing on climate change, otherwise it will always be a recipient of larger level actions and discussions. It is obvious that both Gujarat and also to an extent India can do limited things in a global phenomenon. The other more limited scope is to take action yourself. Here two approaches are there. One is somewhat bullish in terms of physical investments like structures to stop sea erosion. These have limited possibilities since impacts can be elsewhere close by. The other is building institutions which are aware and look at alternative possibilities around and at national and larger levels. Difficult but worth trying as alternates do not exist.

This book has opened many lines of enquiry for Gujarat which is an ecologically fragile region. The more we understand that and build it into our policies, the better-off we will be.

—Yoginder K. Alagh
Chancellor,
Central University of Gujarat,
Gandhinagar.

Preface

Gujarat is disaster prone state. Recent climate change threats have increased the frequency and severity of climate induced disasters. In order to act proactively for disaster management, it is necessary for the state to have an idea about the potential risk due to major hazards. Identification of potential risk can help state decision-makers to take their decision of intervention to reduce the risk and mitigate an adverse impact of disasters in advance. It also essentially helps in prioritising the areas of intervention where disaster mitigation activities need to be more intensive. Further, it ultimately facilitates the governance system in the process of an efficient allocation of financial resources at the time of disaster.

Vulnerability measurement model derived in this study has essentially identified the drivers of vulnerability prevailing in the state. The pattern of vulnerability changes over time and space. Therefore, it was important to explore the method of identifying the drivers of development. This study, in the regards, establishes the methodological approach to identify the most appropriate drivers of vulnerability. Such an effort is helpful to assess the prevailing pattern of vulnerability at any point of time and region. This study does not intend to project the scenarios of vulnerability towards climate change. However, it provides the baseline scenario on which different scenarios can be projected.

Disaster management in Gujarat incorporates risk reduction strategies to deal with risk due to various disasters. Based on this study, the most effective disaster risk reduction strategies have been identified. Such strategies can be blended in ongoing development programmes and would enable them to address the disaster risk in the state.

The study has particularly focused on identifying districts which have very high potential risk of severe effects due to climate-induced disasters. By ranking the districts as per the risk which is calculated comprehensively, the study will help policy-makers in prioritising their interventions for proactive disaster management strategies. This has long lasting implications in terms of effective and efficient resource allocation for disaster risk reduction. Such an effort also reduces the regional disparity in development pattern existing within the state.

This monograph is the outcome of constant support and encouragement by several individuals. First and foremost, I would like to express my sincere acknowledgement to Prof Y.K. Alagh for his valuable inputs and encouragement. A special word of thanks is due to Prof Niti Mehta for her constant support. This study has benefited a lot from discussions with many experts and officials.

I am thankful to Mr Pantlu for extending his statistical support as and when needed. I also acknowledge the help rendered by Mr Satisan and Mr Ashwin Rawal. I would also like to extend my thanks to Mr Kirit Shah for his good administrative support. I would like to express my sincere thanks to ICSSR for giving me this opportunity to undertake this study. Last but not the least, I extend my gratitude to vulnerable community, who made this study possible by giving their valuable responses.

— Shital Shukla

1 Introduction

Overview

One of major implications of climate change is seen in terms of increase in frequency and intensity of extreme events [Intergovernmental Panel on Climate Change (IPCC), 2007]. These extreme events are interactive, dynamic and functional in nature. As these events interact with human system, they turn into natural disasters. Extreme events occur on various scales. They also interact with different levels of development of a human system. As a result, the magnitude of an adverse impact of extreme events on society varies according to degree of interaction. Therefore, these events are dynamic in characteristic. It is also obvious that these extreme events are functional because the scale and nature of these events are partially determined by human activities. When such extreme events take place on inhabitant area, their impact turns into natural disasters or a catastrophe with huge socioeconomic losses. Such losses are setback for development process. Thus disasters prevent the path of development.

Due to its distinct characteristics, these extreme events are called climate-induced disasters. The observation made by the Centre for Research on the Epidemiology of Disasters (CRED) under international disaster database very clearly indicates that climate-induced disaster in Gujarat state has increased more than three times during the last two decades. These disasters pose a risk of potential losses. However, the risk is a function of not only size and magnitude of these events but also of vulnerability existing in a society. In response to such recognition, there has been a paradigm shift in the recent policy development in the context of disaster management. Disaster management is an execution of a strategy to deal with disasters using available resources efficiently and effectively. Disaster management is now no more only relief and response activities but it has included pro-active strategies like preparedness, mitigation and development of various resources. In fact, disaster management is considered as a long-term process which includes various activities such as prevention, mitigation, preparedness, response, relief and rehabilitation. These activities need to be carried out in three different phases, namely; pre-disaster, during disaster and post-disaster. In this way, disaster management approach needs to be developed in a comprehensive way.

A strong need has emerged in the state today for mainstreaming disaster management into development goals. Mainstreaming disaster management into developmental goals means systematic and more conscious ways of integrating disaster risk reduction into development process. In other words, one needs to see that the process of development and the kind of development choices do not create or increase the risk of disasters. This has reflected in the recognition that disaster risk reduction strategies must be addressed as one of the developmental goals. Therefore, it is of an immediate necessity to develop tools which help effectively in mainstreaming disaster risk reduction strategies into developmental goals.

In this context, measuring vulnerability and risk due to climate-induced disasters is the first necessary step for any serious contemplation of disaster reduction strategies of climate change policies. Measuring vulnerability risk is an emerging technique which can systematically assess and analyse vulnerability and risk due to future possible climatic hazards. It can reveal many adverse effects which are led by climate-induced disasters on society. Therefore, it helps state as well as society for preparing and adopting adequate and successful measures for reducing the risk factors. In other words, measuring vulnerability and risk can show direction for preventing losses. It leads towards preparedness to reduce recovery time rather than dependence on relief and reconstruction. In short, measuring vulnerability and risk can help decision makers in prioritising their intervention for pro-active disaster mitigation strategies and climate change policies.

Gujarat is one of the most disaster prone states in India. The state faces various climate-induced disasters frequently and suffers from their adverse impact. Recently, the state has taken many steps to cope, prevent and mitigate the adverse impact of climatic hazards. The major efforts include the passing of disaster act, formulating disaster policy and implementing it through the establishment of a nodal agency called Gujarat State Disaster

Management Agency (GSDMA). It has also established an institute called Gujarat Institute of Disaster Management (GIDM) to provide various training programmes in order to strengthen capacity building. The state has also carried out district and *taluka* level disaster management plans.

The state has prepared vulnerability maps in order to implement disaster management strategies very systematically. These maps considered physical vulnerability as the major criteria affecting the degree of vulnerability towards disasters. Nonetheless, vulnerability is a dynamic and complex concept. It is, therefore, important to conceptualise vulnerability before measuring it. In this context, identification of drivers of vulnerability which causes an unequal degree of risk among spatial units is an important exercise.

As mentioned earlier, the risk due the disasters is not only a function of frequency and severity of climatic disasters but also equally a function of vulnerability of the elements, which are exposed to hazards. It is now well accepted that vulnerability is an extent to which disaster may damage or harm a system. It depends on system's sensitivity and ability to adapt to new conditions. Therefore, it is required to measure prevailing vulnerability in the state and also to assess the future potential risk due to various climate-induced disasters, which frequently affect the state. This takes into account two major contributing sets of factors, namely probability of climatic disasters and different dimensions of vulnerability existing in the state. Measuring probability of climatic disasters requires technical assessment. Hence, it is a subject of applied science. However, measuring vulnerability is a subject of social science. Such measurement requires a systematic methodology and approach. It is, therefore, essential to understand the concept and its emergence of vulnerability in different school of thoughts.

Vulnerability Concept: Emergence and Development

The root of vulnerability concept lies in the studies of natural disasters. Later on, the development of the concept has taken placed in both applied and social sciences simultaneously. Though it is difficult to pinpoint any precise definition of vulnerability, it is overtly defined in applied and social sciences.

Vulnerability by its simple definition refers to the susceptibility of a person, group, society, region or system to any external risk. There are a number of attempts made to define vulnerability, they all lead to the simple meaning explained in above definition. Although it is not possible to define vulnerability precisely, the concept of vulnerability provides flexibility to accommodate a number of phenomena under it. Therefore, the term vulnerability is relative and explanation of the term depends upon various factors. It is a state of vulnerable which refers to the state of perilous situation. Some of its synonymous idioms provide more clarity in understanding the meaning of vulnerability. Idioms such as between two fires or between wind and water reflect the state of vulnerability more evidently in which the object is stuck up in a precarious position, exposed or unprotected and defenceless. In other words, the circumstances in which an object is trapped between an external risk and internal weaknesses, which make it more prone to get affected.

Vulnerability concept possesses multidisciplinary dimensions. There are three major schools of thoughts that have contributed to the development of a conceptual framework of vulnerability. These are social science stream, disaster stream and global environmental change stream. The contribution of social science stream has remained significant in theoretical development of the concept. Disaster stream looks at the vulnerability as a part of disaster risk. However, global environmental change stream considers vulnerability as an expression of unhealthy man and environment relationship. The following is the elaborate review of contribution from these three schools of thoughts.

Disaster Stream Models of Vulnerability

Disaster school of thought characterises vulnerability as a factor of hazard and risk. It considers vulnerability as one of the contributing factors towards the risk. This discipline separates the vulnerability components from other components such as hazard, exposure and capacity. The risk is a sum of all these separate factors, namely hazard, exposure, vulnerability and capacity (Davidson, 1997 and Bollin *et al.*, 2003). Hazard is measured in terms of probability and severity and exposure is measured in terms of infrastructure, population and economy. Although vulnerability here considers physical, social, economic and environmental aspects, it does not capture the parameters explaining specifically human vulnerability. Capacity and measures include physical planning, social capacity, economic capacity and management of disaster risk at macro level.

Vulnerability in a framework of disaster risk is represented mainly by physical infrastructural and demographic indicators. Such framework overlooks the social aspects of vulnerability. Most of these conceptual frameworks are developed on the basis of earthquake risk and therefore focus more on physical vulnerability. Going beyond the summation of separate components, Villagran de Leon (2006) developed a triangle of risk by depicting the three angles with hazard, vulnerability and deficiency in preparation. He considered an exposure as a part of hazard component and vulnerability as a degree of fragileness of the system. His model recognises deficiencies in preparedness as one of the important separate factors increasing the risk. However, in reality, the vulnerable situation may result into deficiency in preparation or vice versa. This fact is ignored in models developed by disaster stream. In fact, all the separate components presented by these models are interdependent and upshot in an overlapping effects. Thus, the triangle risk framework is one step forward to define vulnerability in a hazard risk framework which considers vulnerability as a pre-existing situation.

Blaikie in his pressure and release model for the first time argued that the risk is not only a function of natural hazards but also a function of spatial and temporal exposure of a society exemplified by a varying degree of vulnerability towards extreme events (Blaikie *et al.*, 1996). Wisner developed this model further and explained how disaster happens when hazards interact with vulnerability (Wisner *et al.*, 2004). They see that disasters are the ways to release the pressure created by vulnerability. Vulnerability in this sense is emerged as a process of three stages, namely root causes, dynamic pressures and unsafe conditions. Inequality in accessing and unequal distribution of the resources, opportunities and power is the root cause of creation of vulnerable situation. When these causes are exposed to dynamic processes of development, they sometimes result into unsafe conditions. Such unsafe conditions increase the vulnerability towards natural hazards. This model goes beyond the identification of vulnerability and explores underlying forces contributing to the vulnerable situation. Addressing vulnerability in this approach recommends for addressing the ongoing economic, political and social systems to minimise their ill effects on society or a region. The framework is very useful at all levels not only in understanding the concept of vulnerability but also for measuring the degree of vulnerability.

With regard to the application of vulnerability into risk management, the United Nations adopted a disaster-based approach in their International Strategy for Disaster Risk Reduction framework (UN/ISDR, 2004). The framework regards vulnerability as an important determinant of risk and can be classified into different components such as social, economic, physical and environmental vulnerability. It further underlines the fact that the vulnerability assessment is the most important exercise for disaster risk reduction. Vulnerability-based approach is the most common approach to identify the appropriate measures to reduce the risk.

To put it candidly, the disaster stream has pinned the vulnerability concept to the hazard analysis particularly while analysing the risk and formulating risk reduction strategies. The Inter-American Development Bank has contributed significantly towards developing a variety of indices to measure vulnerability by using disaster-based approach.

Social Science Framework of Vulnerability

Social science theories and models have tried to provide various frameworks to understand the vulnerability more clearly. It is worth mentioning here the work contributed by Bohle *et al.*, 1994. Bohle worked extensively on vulnerability towards severe droughts. From his experience of drought-related works, he defines vulnerability as an outcome of interaction between coping capacity of community and their exposure towards hazards. Bohle explicated vulnerability in two different dimensions, namely internal and external, and thus his work revealed double structure of vulnerability. The internal dimension of vulnerability represents the capability of the community or society or a region to cope with, resist and recover from the impact of hazards. However, the external dimension of vulnerability relates to the exposure of such community or society or a region towards any stress or risk. It is the first time that exposure was considered as a part or component of vulnerability. Such considerations make the social science stream different from the disaster stream which looks exposure as a component of risk not the vulnerability. Further, the exposure as an external dimension of vulnerability in social science stream encompasses not only the physical exposure but also deliberates demographic, economic and environmental exposures. In fact, Bohle has included even social and institutional features of exposure which enhance vulnerability.

Concisely, Bohle's double structure of vulnerability has considered the vulnerability as a result of external shocks and weak coping capacity as an internal side weakness. Interestingly, the vulnerability defined by Bohle over social space considered political, economic and institutional capabilities of people. It is worth noting that the external factors are considered as a part of vulnerability which is contradictory to theories of vulnerability in disaster stream.

As far as internal dimension is concerned, Bohle incorporated coping, responding and recovering capability of a community/society or a region. Although he has not elaborated internal dimensions of vulnerability, there are a number of factors which determine the coping, responding and recovery capabilities. Such factors are well explained in a sustainable livelihood framework provided by Robert Chambers (Chambers and Conway, 1992). The sustainable livelihood framework explores the ways and means for improving the capability of a community to cope with and recover from external shocks as well as for maintaining livelihood and natural resource base during the time of any external shock or stress. This is how the framework explores the vulnerable features of the community/society or a region. The concept of sustainable livelihood emphasises on how various types of capitals, namely human, natural, financial, social and physical can be converted into livelihood assets under various circumstances. Transformation of such capitals under various structural processes influence the ability of poor in maintaining and protecting their means of livelihoods not only at the time of external stress and shocks but also afterwards in a sustainable manner. This framework is very useful and can be extended to study livelihood vulnerability towards natural disasters. Such efforts include the strategies for enhancing capability of poor to reduce their vulnerability. Various structural processes at macro level include governance system, private sector participation and other religious and cultural processes that determine the access to and impact on various livelihood assets in the context of vulnerability concept. This model builds strategies for reducing vulnerability and accentuates the decentralised system of governance which uses a bottom-to-top approach for decision making.

In this context, two models, which have contributed significantly to explain vulnerability in terms of risk measurements, are developed by the United Nations University. They are onion model and Bogardi, Birkmann and Cardona (BBC) conceptual framework. The onion model is based on an extensive research on flood. It presumes that the risk due to hazards can affect different spheres of vulnerability such as economic and social. Further, the social sphere has also diversity within it as per preparedness and coping capacity. Thus, when hazard occurs it affects the outer sphere in term of economic losses. If the inner spheres of social systems are affected, the hazard turns into disasters. The model underlines the concept of social vulnerability as an important dimension of potential losses due to disasters.

BBC framework developed by Bogardi and Birkmann (2004) and Cardona (1999; 2004) attempts to address vulnerability issues discussed in all the above models. The holistic approach provided by them puts vulnerability in a specific context of sustainable development and human security. It underlines the need to focus on reducing vulnerability rather than carrying out risk analysis. It strongly puts forward the fact that reducing vulnerability can help in sustaining development and achieving human security. It adopts three pillars of sustainable development as dimensions of vulnerability. These are economic, social and environmental aspects of vulnerability. It emphasises on proactive approach for reducing risk and vulnerability. The model incorporates sustainable livelihood frameworks for improving various assets in order to reduce vulnerability. Environmental focus in this approach reveals the fact that disasters sometimes emerge out of the environmental negligence and at the same time, it leads to further environmental degradation. The most important aspect of this model is that it considered exposure as an integral part of vulnerability. This is because the model presumes that the exposure increases the vulnerability towards hazardous events. The present study has adopted this fact from BBC model and has presumed that exposure is a part of vulnerability rather than hazard. This is because it can capture the vulnerability in its true sense. For instance, equally vulnerable systems can face different risk due to the difference in exposure towards risk and this can change the degree of vulnerability of a system towards any external stress.

Global Environmental Change School of Thought

Presently, a lot of work is being carried out under this school of thoughts. Recent knowledge on global warming and resultant climate change threats has contributed lot in the development of this school of thought. However, the work in this school of thought has way back from the significant work contributed by Turner *et al.* (2003). Global environmental change school of thought looks at the vulnerability from the view point of man–environment relationship. This stream also analyses the concept of sensitivity and resilience in the context of vulnerability. The vulnerability analysis in this stream mainly carried out to formulate suitable adaptation strategies at various levels. It is important here to mention Turner's model of spatial vulnerability (Turner *et al.*, 2003). Turner developed a vulnerable framework focusing on a space-based approach which captures the coupled human and environmental characteristics of any region. It considers human and environment subsystem as parts of one system. The framework includes the inter-linkages of human and environmental subsystems as a major focus. It believes that the response in human subsystem can make bio-physical subsystem more or less vulnerable or vice versa. This is very important from the sustainable development points of view. The strong sustainability concept pinpoints on keeping each and every capital intact and does not favour for replacing one capital in place of others. Strong sustainability followers try to protect the natural capital to maximum extent. Turner's contribution is extremely important in supporting strong sustainability because it does not focus only on a concerned system but considers the coupled subsystems and linkages among them. In short, Turner's framework pens down the importance of considering human and environmental system jointly in his study. The present research work has adopted this fact to develop a regional model and has also considered human and environmental capitals jointly in analysing regional vulnerability.

Application of the Vulnerability Concept in Risk Measurement: Recent Approaches

Evolution of conceptual models has resulted into many attempts for measuring vulnerability at regional level. Many regional level models were developed by identifying vulnerability indicators for the assessment of vulnerability towards external stressors at micro level. Such efforts were carried out both in natural and social sciences. In social science, the emphasis was given on resilience which is inversely proportional to vulnerability. Subsequently, the same approach was used in mapping vulnerability to climate variability and climate change.

The development of regional model on natural hazards as well as climatic variability has its base on a famous model known as hazard-of-place which was developed by Cutter in United States. The original work of Cutter used the hazard-of-place approach which was later on, extended to measure resilience of community to natural hazards (Cutter *et al.*, 2008). This approach developed resilience indicators for measuring community vulnerability. Similar approach was used to map vulnerability at country level by O'Brien *et al.* (2004). He specifically applied this model to measure vulnerability of agricultural sector in India. His study mapped the vulnerability of agricultural sector to two external stressors, namely climate change and globalisation. He assumed that the sector is doubly exposed to various stresses. This makes the sector highly vulnerable to climate change.

IPCC added a new methodology to measure vulnerability by identifying the indicators of exposure, sensitivity and adaptive capacity. IPCC methodology is mainly used to measure vulnerability of a particular sector such as agriculture, water, coastal area, etc. towards climatic stressors. Many scholars in India applied this methodology to measure vulnerability of a particular sector towards climatic variability or climatic hazards. It is worth noting here the study carried out by Ravindranath *et al.* (2011) on preparing climate change vulnerability profile for North East India. The study used an index-based approach in which indicators representing vulnerability of various sectors were identified by using statistical techniques. Similar study carried out by Mishra *et al.* (2011) focused on assessing role of contributory factors and their interplay towards the vulnerability of agriculture sector in Orissa. The former study considered climate variability as a climate stressor while the latter one used extreme events as an external stressor. As both the study focused on agriculture, various indicators of precipitation were used to describe external factors. Both the studies mapped vulnerability of agricultural sectors geographically by using maps.

At national level, vulnerability atlas of India prepared by the Building Materials and Technology Promotion Council has studied district level vulnerability in India (GoI, 2005). However, their approach focused on physical aspect of vulnerability. The most important indicator for assessing vulnerability at district level was housing stock vulnerability. The major objective of the atlas remained to reduce loss of lives, property damage and economic disruption. Following the similar methodology, Gujarat State Disaster Management Agency has prepared a vulnerability atlas at state level. Such atlas provided data on districts which are more vulnerable to various hazards. Such a study provides information on the districts where probability of particular hazard is more. However, the impact of disasters does not depend only on probability and intensity but it also depends on the elements of vulnerability towards various disasters. Therefore, it is important to derive these elements of vulnerability. In order to identify such elements, it is important to understand the application of this concept.

One of the important contributions to the assessment of vulnerability has been carried out by Anu Kapur (2010) in her book entitled *Vulnerable India*. She argues that disasters are not natural events but human suffering due to neglect and apathy. Emphasising on voice of concern towards the conditions of living that makes people assailable to natural forces of various forms and magnitude, she puts human at the core of disasters' studies. The work carried out by her identified three major components of vulnerability, namely disadvantaged people, fragile living and lacking services. Her study further developed the vulnerability indicators and found that the coastal region of India was the most vulnerable region compared to other parts of India. Gujarat having the largest coast line is the most vulnerable state in India. Therefore, the state needs a special focus on research related to vulnerability. This study has made an effort to work on such issues more comprehensively.

In short, disaster risk and vulnerability analysis is relatively new area of research and has received attention for the last few decades. The most prominent and land mark study in this area is carried out by the Centre for Hazards and Risk Research at Columbia University in collaboration with the World Bank. The study entitled 'Natural Disaster Hotspots: A Global Risk Analysis' is the first attempt to assess the risk potential at global level (Dilley *et al.*, 2005). This study has assessed the potential risk in terms of two major outcome indicators of disasters, namely mortality losses and economic losses. It has followed a methodology developed in the above-mentioned study. However, it has modified the methodology and selected the indicators which are appropriate for state-level analysis. Further, the current study has gone beyond and assessed the social losses by developing vulnerability indicators at district level.

The characteristic of vulnerability depends upon two broad categories of factors, namely internal and external factors. First set of factors, namely internal factors, comprises of internal characteristics of a system such as society, community, region, person or group. Another set of factors on which the vulnerability depends is called external factors. External factors or risk can also provide the information regarding in which discipline the concept of vulnerability has emerged. For instance, the hazard as an external risk or factor is usually considered in disaster stream. In a social science stream, the external factors are seen in terms of external shocks such as hazards, global economic crisis, internal economic crisis, market failure, etc. The man and environment stream looks at the external factor in terms of climate change and other environmental threats. This study has assessed the internal elements of vulnerability which are important determinants of prevailing status of vulnerability in the state.

Objectives of the Study

The main aim of this study is to assess the risk and vulnerability of climate-induced disasters. The specific objectives included:

1. To assess the impact of disasters on society and economy of Gujarat.
2. To assess the disaster management process in the state
3. To identify determinants of prevailing vulnerability in the state
4. To prepare risk index incorporating prevailing vulnerability in the state for all the three selected climate-induced disasters.

In order to fulfill the above objectives, this research has made an effort to develop a state-level inventory of major climate-induced disasters and assessed the potential risk variability and the degree of vulnerability among various parts of the state. It has particularly focused on identifying districts which are highly vulnerable due to climate-induced disasters. An overall risk is measured in terms of three disaster-related outcomes, namely mortality losses, social losses and economic losses. Further, the present study has carried out district-level analysis of potential risks due to major climate-induced disasters. By identifying the extreme vulnerable districts, the study intends to help policy makers in prioritising their interventions for proactive disaster management strategies

Implications of the Study

Gujarat is a disaster prone state. Recent climate change threats have increased the frequency and severity of climate-induced disasters. In order to act proactively for disaster management, it is necessary for the state to have an idea about the potential risk due to major hazards. Identification of potential risk can help state decision makers to take their decision of intervention to reduce the risk and mitigate an adverse impact of disasters in advance. It also essentially helps in prioritising the areas of intervention where disaster mitigation activities need to be more intensive. Further, it ultimately facilitates the governance system in the process of an efficient allocation of financial resources at the time of disaster.

Vulnerability measurement model derived in this study has meaningfully identified the drivers of vulnerability prevailing in the state. The pattern of vulnerability changes over time and space. Therefore, it is important to explore the method of identifying the drivers of development. This study, in this regard, establishes the methodological approach to identify the most appropriate drivers of vulnerability. Such an effort is helpful to assess the prevailing pattern of vulnerability at any point of time and region. This study does not intend to project the scenarios of vulnerability towards climate change. However, it provides the baseline scenario on which different scenarios can be projected.

Disaster management in Gujarat incorporates risk reduction strategies to deal with risk due to various disasters. Based on this study, the most effective disaster risk reduction strategies can be identified. Such strategies can be blended in ongoing development programmes. This would enable ongoing development programmes to address the disaster risk in the state.

Briefly, it has particularly focused on identifying districts which have very high potential risk of severe effects due to climate-induced disasters. By ranking the districts as per the risk which is calculated comprehensively, the study has an intention to help policy makers in prioritising their interventions for proactive disaster management strategies. This has long lasting implications in terms of effective and efficient resource allocation for disaster risk reduction. Such an effort also reduces the regional disparity in development pattern existing within the state.

This book has six chapters. The first chapter introduces the subject and elucidates the need for carrying out the study. It further illuminates the concept of vulnerability from its emergence to development in various disciplines. It also clarifies the objectives and implications of the study.

The second chapter begins with defining and classifying the notion of disaster. It further assesses the risk due to three climate-induced disasters which are faced by the state. The assessment of risk is carried out by analysing historical trend of natural disasters and their social, economic and macroeconomic effects on state economy. It has further looked into the state position in responding to such impact

The third chapter assesses the existing status of development in the state of Gujarat. It argues that disasters and development are mutually related. Disasters can create a setback in a process of development. At the same time, the type of development can increase or reduce the risk due to disaster. The chapter explains different dimensions of development. Various indicators were analysed to assess the demographic, socieconomic and environmental aspects of development.

The fourth chapter has attempted to identify the vulnerability indicators at state level. It further goes beyond identifying the indicators of vulnerability and also endeavours to quantify those indicators at district level. It then explores a regional model for assessing the pattern of vulnerability within the state and identifying the determinants of vulnerability existing in the state. Later on, the vulnerability assessment is applied to measure the risk due to climate-induced disasters. Such an exercise is useful for identifying the most vulnerable district in the state for regional planning.

The fifth chapter presents several case studies to assess vulnerability at micro level. The vulnerability analysis at the micro level was carried out to serve two purposes. The first one was to ensure that whether the macro level vulnerability drivers are valid at micro level. The second purpose was to understand the vulnerability pattern existing at village level and addressing such vulnerability through development planning. The case studies have examined that how poor socioeconomic status and environmental degradation can make the region more vulnerable towards climate-induced disasters. In this way, these case studies helped in understanding the relationship of climatic variability risk and vulnerability of the region.

The last chapter shows ways and means of integrating risk reduction strategies into development planning.

2 Disaster Profile of Gujarat

Trend and Impact Assessment

Gujarat is a highly disaster-prone state in India. Having a longest coastline among all the states in India, the state is facing various disasters over a period of time. Such disasters can prevent the development process by diverting funds from other planning departments. Therefore, it is very important to understand the risk due to disasters in the state of Gujarat. This chapter begins with defining and classifying the notion of disaster. It further assesses the risk due to three climate-induced disasters which are faced by the state. The assessment of risk is carried out by analysing historical trend of natural disasters and their social, economic and macroeconomic effects on state economy. It has further looked into the state position in responding to such impact.

Disaster: Meaning and Classification

In a literary meaning, disaster is an occurrence causing widespread destruction and distress. A more intense definition of disaster states that a disaster is a natural or man-made (or technological) hazard that has come to fruition, resulting in an event of substantial extent causing significant physical damage or destruction, loss of life or drastic change to the environment. Gujarat state government Act defines disaster as an actual or possible likely event that leads to loss of human life, injury or illness, loss of movable and immovable property and degradation of environment [Government of Gujarat (GoG), 2003].

Disasters are classified in various ways. The basic traditional classification of disaster distinguishes natural and man-made disaster separately. Disasters are complex phenomena and sometimes emerge out as a result of interaction between natural causes and human interruption. Hence, disasters are also classified in three categories such as natural, anthropogenic non-intentional and anthropogenic intentional. Natural disasters originally could be categorised from their way of origin. The Federal Emergency Management Authority in the United States puts natural disasters into various categories such as atmospheric or meteorological, geological, hydrological, extraterrestrial and biological. Meteorologically originated disasters are tropical cyclone, thunderstorm, tornados, lightning, hailstorms, windstorms, ice storms, snowstorms, blizzards, cold wave, hot wave, avalanches, fog and frost. Geological disasters are origin of geological movement of earth and include earthquakes, volcano, tsunami, landslide, subsidence, mudflow and sinkholes. Hydrological disasters are floods, droughts and wildfires. Meteorites and asteroids are originated from extraterrestrial processes. Diseases, epidemic, pandemics (epidemic of infectious diseases through human population), overpopulation and famine are resultant of biological process.

In India, diverse classification systems for disasters exist. An important classification carried out by High power committee (GoI, 2001) identified 31 disasters and classified them into five different groups namely, water and climate related disasters, geological disasters, chemical, industrial or nuclear disasters, biological disasters and accident related disasters. Water and climate related disasters included floods and drainage, cyclones, tornadoes and hurricanes, hail storms, cloud bursts, heat wave and cold wave, snow avalanches, droughts, sea erosion and thunder and lightning.

This study has used International Database (EM-DAT) series. EM-DAT series have been prepared by the Centre for Research on Epidemiology of Disasters (CRED). EM-DAT makes a distinction between natural and man-made disasters. As the focus of this study remains on climate-induced natural disasters, cyclone, flood and drought are taken into consideration. It is important to note here that the natural disasters are so complex that it is not easy to distinguish them with man-made disasters. In fact, Delaware school of thought has stated that all disasters are the result of human failure to introduce appropriate disaster management measured or in other words, the risk due to disaster is equally a function of inappropriately managed risk. EM-DAT series further classifies the natural disaster into geophysical (earthquake), meteorological (cyclone), hydrological (flood), climatological (drought) and biological (epidemic). Most prominent climate-induced disasters faced by Gujarat are cyclone, flood and drought. Further, these disasters are classified according to their pace of occurrence. Drought is slow onset disasters, while flood and cyclones are rapid onset disasters.

Assessing the Risk Due to Disasters: A Methodological Approach

Risk is a function of hazards and vulnerability. Therefore, it is very important to understand the pattern of hazard as well as its socioeconomic impact. As Gujarat has suffered from various climate-induced disasters, there exist a lot of data on impact of various disasters. Basically, the risk due to disasters in this chapter is assessed by assessing the historical trend of disasters and its various impacts captured in the following framework.

1. *Probability of Disasters:* The probability of hazards is analysed by assessing the historical trend of frequency and significance of various disasters faced by Gujarat. Laplase's formula of probability is used to estimate the probability of various climate-induced disasters.
2. *Social Impact*: The social impact can be expressed in terms of direct impacts like number of persons killed, injured and affected as well as indirect impacts like spreading of diseases or other health impact in long term. In this study, the social impact is assessed by looking into the extent of death toll, injury, homelessness and disaster-affected people.
3. *Economic Impact*: Economic impact is divided into two types of impact, direct and indirect. A direct impact includes the damage estimation of assets. An indirect impact is assessed through the losses of production output and disturbance in various services as a result of economic destruction which finally leads towards the macroeconomic impact. Further fund diverted to relief activities from other development expenditure leads to high fiscal deficit in the state budget. The study has estimated the direct economic losses and also tried to explain the macroeconomic impacts of disasters on the state economy.

Further, it is assumed here that risk due to climate-induced disaster depends on probability and intensity of such extreme events. As far as intensity is concerned, it can be estimated by assessing an impact of such disasters. Three different indicators of impact are derived for assessing intensity of the climatic disaster. These are coefficient of mortality, coefficient of social vulnerability and coefficient of economic vulnerability. Methodology for calculating these coefficients is adopted from the United Nations Development Programme's (UNDP's) approach. Coefficient of mortality is derived by calculating total death toll divided by total population of the state. The 2001 data have been selected as the base year. Coefficient of social vulnerability has been derived by dividing total affected people with total population of the base year. Coefficient of economic vulnerability is derived by dividing total damage with gross domestic product of the selected base year. In short, risk is measured in terms of three types of disaster losses, namely mortality, social and economic. It has considered three major climatic hazards, namely flood, cyclone and drought. The calculated risk is not an absolute one but relative risk.

The following are the steps for analysis:

1. An average mortality, social and economic losses for all the three climate-induced disasters are calculated by aggregating data of last 30 years. These are denoted as Mh (average death toll), Sh (average number of people affected) and Eh (average economic loss) for h type of hazard.
2. Coefficient of mortality, social and economic vulnerability is calculated as

   Mrh= Mh/State Population of year 2000

   Srh = Sh/State population of year 2000

   Erh = Eh/Net State Domestic Product (NSDP) of Gujarat of the year 2000

Finally, the present scenario of expenditure made on various disasters by the state government is also assessed from the perspective of disaster management.

Data Sources

Data on disasters in the state are available with various departments such as relief commissionarate, agricultural departments. However, such information remains in piecemeal manner and cannot provide any useful analysis. The Gujarat State Disaster Management Authority (GSDMA), which is supposed to act as a data repository agency, has published data on disasters on its website. These data are collected from the University of Colorado. In fact, the University of Colorado is a client of EM-DAT and therefore uses the data from EM-DAT series which have been prepared by the CRED at Brussels. Therefore, this study has also used data on disasters and their impact published by the International Disaster Database from EM-DAT series.

An up to date inventory on frequency and impact of climate-induced disasters has been prepared by using data from multiple sources. The major source has remained EM-DAT a global disaster database, Department of Relief Commissioner, Indian Meteorological Department, Geological Survey of India and the Central Water Commission.

EM-DAT series provides statistics on number of occurrences and their impact in terms of death tolls, affected people and damages. EM-DAT series has defined disaster as an unforeseen and often sudden event that causes great damage, destruction and human suffering. The data set considers disaster in its list only when fulfilling any one of the following criteria:

1. Ten or more people are reported killed.
2. Hundred or more people reported affected.
3. Declaration of a state of emergency, or
4. Call for international assistance.

Probability of Disasters: Frequency and Significance

With the longest coastline in India and a variety of geological formation, Gujarat is prone to many hazards. The intensity and frequency of hazards in Gujarat shows that the state has faced three major climate-induced extreme events (hazards) which turn into the disasters. These three types of significant disasters are drought, flood and cyclone.

Table 2.1 shows the frequency of disasters faced by Gujarat over a period of time. It indicates that the frequency of disasters has increased sharply over a period of time indicating the increase in risk and potential impact of disaster on the state.

Table 2.1

Frequency of Disaster Over a Period of Time as well as by Type of Disasters

Year	*No. of Events*
1930s	1
1940s	3
1950s	3
1960s	3
1970s	5
1980s	10
1990s	16
2001-03	7
Total	48

Source: EM-DAT: The Office of U.S. Foreign Disaster Assistance (OFDA)/CRED International Disaster Database.

An annual probability for each disaster has been calculated with the classical approach of probability calculation. The probability calculation has assumed that the chances of occurrence are likely to be equal for each of the five significant disasters (Gupta, 1969). Laplase's formula of probability in a classical approach is the ratio of the number of favourable cases to the total number of equally likely cases. Thus the probability of occurrence of X disaster is

$$P(X) = \frac{a(X)}{N},$$

where a(X) is the number of occurrences of X disaster,

N = Total number of years considered.

The probability is then converted into percentage by multiplying it with hundred. It is clear from Table 2.2 that the annual probability of happening any disaster in the state is 0.50. In other words, the state faces disaster at every alternate year. The probabilities of various natural climatic disasters differ in the state. Table 2.2 presents the annual probability of three major climatic disasters faced by the state.

It is interesting to note here that an annual probability of flood is the highest among all the three selected climatic disasters. This is followed by cyclone and drought with probabilities of 17.5 per cent and 7.5 per cent respectively. Drought is also a very frequent phenomenon in the state with arid and semiarid climate. However it is not considered as disasters when there is partial drought. In fact, one of the estimates in Gujarat shows that the state faces drought in every three years. Having a quite high annual probability, the flood causes high financial burden on the state which ultimately has severe implications on the state economy.

Table 2.2

Annual Probability of Disasters in Gujarat

Causes	*Type of Disasters*	*Frequency of Disasters during last 80 years*	*Annual Probability*	*Annual Probability in %*
	Slow On set			
Climatological	Drought	6	0.075	7.5
	Rapid Onset			
Hydrological	Flood	20	2.500	25
Meteorological	Cyclones	14	0.175	17.5
	Grand Total	40	0.500	50.00

Source: EM-DAT: The OFDA/CRED International Disaster Database.

Social Impact

It is imperative to assess social indicators showing direct and indirect adverse impact of disasters on the state. Three major social indicators showing an adverse impact are number of death toll, number of injured people and number of people who are affected by disasters at a time.

The immediate impact of the disasters is considered by the death toll it created from its impact. The death toll is not only the loss of human lives but also a great psychological loss to their near and dear ones. Table 2.3 indicates the death toll due to various disasters over a period of time. It is clear from the table that disasters have vast social adverse effects. Trends in causalities show that the trend of death due to flood is increasing. It is also important to note here that the frequency of flood in the last two decades has increased. The irrigation department of the state government devotes a substantial proportion of its budget for flood mitigation strategies. Such strategies mainly include structural measures. Preparedness and mitigation actions play a major role in protecting life of local people from flood. The state is in its initial phase as far as risk reduction strategies for flood are concerned.

Table 2.3

Death Tolls Due to Various Disasters that Hit the State Over a Period of Time

Decade	*Slow Onset*	*Rapid Onset*	
	Drought	*Flood*	*Cyclone*
1940s	0	0	56
1950s	0	62	90
1960s	0	879	0
1970s	0	1156	128
1980s	0	1421	1641
1990s	0	2481	3149
2000-03	20	452	5
Total	20	6451	5069

Source: EM-DAT: The OFDA/CRED International Disaster Database

Further, death toll due to cyclone has increased over a period of time. The sudden onset of this disaster does not give chance to protect lives of coastal community. Although there is an early warning system working very well, the local community does not follow the warning mainly because it is a question of survival of their livelihood. It is worth noticing that the state has reported death due to drought. Death due to drought is not very common. Unless and until, there is severe drought, death does not happen. However, the state has reported death due to drought in the last two decades. The death reported due to drought is mainly because of two reasons. Firstly, the heat wave along with drought causes death tolls during the time of drought. Secondly, extreme poverty along with malnutrition during the drought season also results in death. It is worth noticing that the state has reported death due to drought for the first time in the year of 2000. The drought during the year of 2000 was the worst among all, and for the first time, the state reported the first death on its website (http://gujaratonline.com/newsroom/drought). However, the state did not report total death toll. There are other sources which also reported deaths due to drought in Gujarat. However, this study has used only the CRED source of data for comparison.

Table 2.4

Number of People Injured Due to Various Disasters that Hit the State Over a Period of Time

Decade	*Number of People Injured*			
	Drought	*Flood*	*Cyclone*	*Total*
1940s	0	0	10	10
1950s	0	0	0	219
1960s	0	0	0	0
1970s	0	23	100	123
1980s	0	0	0	0
1990s	0	0	893	893
2000-03	0	0	0	166812
Total	0	23	1003	168057

Source: EM-DAT: The OFDA/CRED International Disaster Database.

Information on injured people is provided in Table 2.4. Injured people are those who suffered from physical injuries, trauma or an illness requiring medical treatment as a direct result of a disaster. With regard to Gujarat, cyclone reported the highest number of people injured when compared to other disasters. The important point to mention here is that the injury in the context of cyclone may be too dangerous and can be sometimes worse than death. Flood has also resulted in injury once during the decade of seventies in Gujarat.

Figure 2.1

People Affected by Climatic Disaster in Gujarat

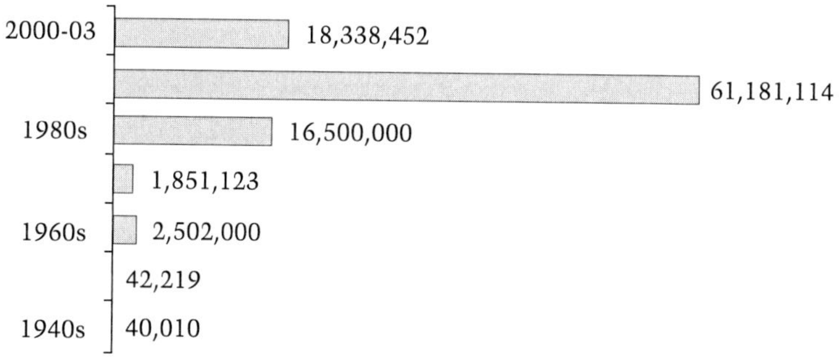

Source: EM-DAT: The OFDA/CRED International Disaster Database.

Figure 2.1 depicting data on people affected by various climatic disasters shows that reported number of people affected by various disasters in different decades has increased sharply over a period of time. Furthermore, the intensity of effect was observed highest in the decade of nineties in which the state faced severe negative social implications of a number of disasters.

Similarly Table 2.5 represents the percentage of population in the state affected annually by various disasters. The table shows that the proportion of people affected by disasters is significant in the decade of nineties. Although the percentages seem to be less, the actual number of people affected is quite high. The proportion of people affected due to flood in the decade of nineties remained 8.8 percentages which alone stands to 3.6 million people in the state. This shows that the actual number of affected people is quite large. For instance, only a single event of flood due to heavy rainfall in 1997 affected 1.2 million people of four districts, namely Ahmedabad, Mehsana, Gandhinagar and Surendranagar (*http:/www.bapscharities.org/ services/ disaster/1990/1997floods.htm*). With regard to the risk due to different disasters, proportion of people affected by flood and drought is higher than those affected by cyclone. The information on affected people indicates that the state has to face either flood or drought almost every alternate year. This shows very clearly that the climatic variability in the state is extremely high. Further, the decade of nineties has experienced all the three kinds of disasters and has faced severe negative social implications in terms of flood, drought and cyclone. One cannot deny that such severe impact is mainly due to climate variability and extremities evolved from anthropological

Table 2.5

Percentage Share of People Affected by Disasters Over a Period of Time

	Drought	Flood	Cyclone
1940s	0	0	0.03
1950s	0	0.025	0.0012
1960s	0	1.21	0.0000
1970s	0	0.60	0.0937
1980s	3.5	0.60	1.3200
1990s	5.13	8.8	0.8331
2000-03	0	2.37	0

Source: EM-DAT: The OFDA/CRED International Disaster Database.

activities. In short, the risk due to disaster is visible during the decade of nineties.

Table 2.6

The Burden of Adverse Impact of Indian Hazards on Gujarat

Major disasters	*The Percentage Share of Gujarat to India*				*Total Affected*
	Killed	*Injured*	*Homeless*	*Affected*	
Drought	2.8	0.0	0.0	36.8	36.8
Flood	25.3	1.6	0.1	35.5	35.0
Cyclone	3.1	5.8	0.0	13.7	12.2
Total	0.5	72.3	7.4	35.1	34.7

Source: EM-DAT: The OFDA/CRED International Disaster Database

Table 2.6 shows that the state's share of total number of people who are affected by major disasters in India is quite significant. In India, about 36.8 per cent of people affected by drought, 35 per cent by flood and 12.2 per cent by cyclones are from Gujarat. In short, in spite of having a well-developed economy, the share of the state in India for total number of affected people (who at least for a time either lost their home, their crops, their animals, their livelihood or their health because of disaster) by climate-induced disaster is more than one forth (34.7%). This indicates that Gujarat is one of the most affected and vulnerable states in India with respect to negative impacts of disasters. All the three major disasters, namely drought, floods and cyclones, put major burden due to their adverse impact on Gujarat.

Economic Impact

As far as an adverse economic impact of disasters on Gujarat is concerned, there are various efforts made for estimating the total loss from various disasters. The EM-DAT series of International Disaster Database carried out jointly by OFDA and CRED provides the data on economic loss only after 1970. The data show that the economic loss due to various disasters has dramatically increased since 1970. The increase in decadal loss from 1970s to 1990s is very high.

Figure 2.2

Damage Due to Disasters

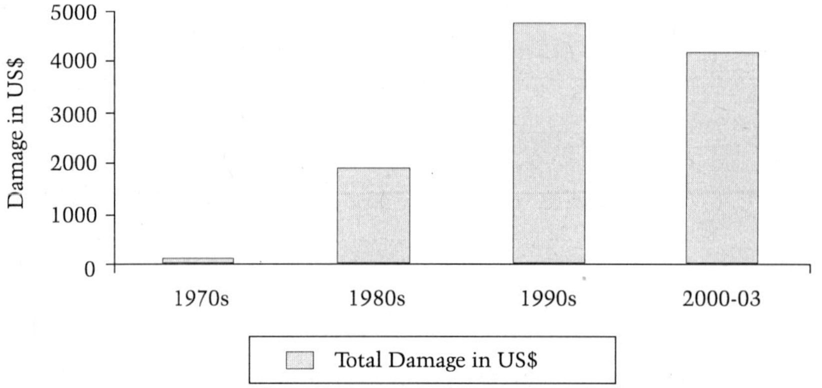

Source: EM-DAT: The OFDA/CRED International Disaster Database, *www.em-dat.net*, Université catholique de Louvain, Brussels Belgium.

Economic loss during 1990s is 30 times higher than that during 1970s. Further, the economic loss during the years of 2001 and 2002 was around US$ 4.2 million which is almost 88 per cent of economic loss due to the damage during 1990s. This indicates the severity of damage is increasing over a period of time in the state. This confirms the fact that the trend of human suffering and economic loss is worsening in Gujarat.

Table 2.7

Economic Impact of Gujarat Disasters

Decadal Years	*Damage Estimation Due to Various Hazards (in US$)**				
	Climate-induced Disasters			*Non-climatic Disaster*	
	Drought	*Flood*	*Cyclone*	*Earthquake*	*Total*
Drought	2.8	0.0	0.0	36.8	36.8
1970s	0	152	1	0	153
1980s	79	0	1810	0	1889
1990s	518	1721	2512	0	4752
2000-03	0	28	0	4169	4197
Grand Total	598	1902	4323	4169	10991

Note: * The figures are deflated on the base year of 2005.

Source: EM-DAT: The OFDA/CRED International Disaster Database.

Table 2.7 compares the economic impact of climatic and non-climatic disasters. Looking from the disaster point of view, Gujarat emerges at global level as an earthquake hit area where earthquake seems to be the most dangerous disaster. Or in other words, Gujarat is always illustrated as an earthquake hit state in the global picture. In fact, the disaster management activities started only after the event of an earthquake. However, this does not mean that other disasters are less dangerous for the state. Comparing the impact of various disasters, cyclones are the worst disasters which have made tremendous economic loss (US$ 4.3 million) in the state. This is followed by earthquake with US$ 4.2 million and flood with US$ 1.9 million of economic losses in the state. The losses calculated by EM-DAT series are very conservative. As per unpublished sources (data prepared for Council of States, 2003), an average economic loss due to flood in Gujarat during 1952 to 2002 has remained US$ 7.26 million, which is extremely high compared to losses observed by EM-DAT series. However, these losses are not comparable mainly because of the unavailability of disaggregated data on economic losses due to flood in Gujarat.

The data provided in Table 2.8 were produced at Council of States in the year of 2003. The statistics provide data of damage due to flood for 50 years. The damage to public utilities is very high when compared to damages to crops and houses. This indicates that the public losses due to flood are quite significant. Another important adverse effect of flood is losses of cattle. This may affect dairy industry which is flourishing in Gujarat.

Table 2.8

Flood Damage in Gujarat (1952-2002)

States/UTs	*Gujarat*
Area Affected (in million hectare)	0.313
Population Affected (in million)	1.846
Damage to Crops Area (in m.ha.)	0.206
Damage to Crops Value (in ₹ crores)	11.12
Damage to Houses (in number)	40253
Damage to Houses value (in crores)	7.095
Cattle Lost Numbers	13260
Human Lives Lost Numbers	139
Damage to Public Utilities (in ₹ crore)	19.549
Total Damages to Crops, Houses and Public Utilities (in ₹ crore)	37.764

Source: *www.gujaratstat.com*

Table 2.9

Significance of Various Disasters in Terms of their Various Adverse Effects

Impact	*India*	*Gujarat*
Frequency of Disasters	Flood, Cyclone and epidemic	Flood and Cyclone
Loss of Life	Drought and epidemic	Earth quake and flood
Injured	Earthquake	Earthquake
Homelessness	Flood and Cyclones	Earthquake
Affected	Drought and Flood	Drought, Flood and Cyclone
Damage	Flood and Cyclone	Cyclone and Earthquake

Source: The author estimated from various sources.

From the above analysis it is easy to portray a picture of disaster significance in terms of their various implications. The most significant natural disasters in India and Gujarat are shown in the above table. Flood, cyclone and epidemic are the most frequent and recurrent disasters in India. Similarly the highest recurrent disasters in Gujarat are flood and cyclone. Nonetheless, earthquake being less frequent is being hostile for the state by affecting a large number of people who died, injured, became homelessness or suffered from damage. Similarly flood, drought and cyclone have also adverse impact on a large number of people who at least for a time either lost their home, their crops, their animals, their livelihood or their health because of disasters.

Macroeconomic Impact

Economic impact of disasters can be classified into direct losses, indirect losses and secondary effects. Direct losses can be captured by estimating the damage in physical infrastructure. The economic impact described in the preceding section includes only direct losses. Such losses are calculated by estimating damage to public stock or assets including housing and commercial buildings, public infrastructure such as public health centres, schools, transport and power sector, water supply and sanitation, communication setup, agricultural assets, etc. Indirect losses are calculated by losses occurred in flow of goods and services due to the direct damage. In other words, losses in production sectors are incorporated in indirect losses. During the disaster years, government has to reallocate the financial resources for relief and reconstruction activities. Direct and indirect losses along with the diversion of plan funds can affect macroeconomic variables such as gross domestic product (GDP), public finance indicators, consumption, inflation, employment, gross savings, capital formation, etc.

It is important to look into the picture that how disasters are affecting the state economy. As mentioned previously, Gujarat has extensively suffered from climate-induced disasters in the last decade. Although it is impossible to capture the macroeconomic impact of each and every disaster, some recent observations on macroeconomic impact of severe disaster could be observed easily.

Let us examine the impact of climate-induced disaster on production sectors. Disasters interrupt the production sectors by damaging assets and disrupting the flow of services and goods.

The impact of disasters on three major sectors is shown in Table 2.10 and Figure 2.3. It is very obvious that the primary sector has shown very high correlation with drought. The production has shown to be reduced during the drought years. In fact, partial drought (drought declared in selected *talukas*) has also shown its adverse impact on agricultural production. Similarly consecutive disasters have affected all the sectors adversely. All the three sectors have shown very sluggish growth during the time of two consecutive droughts of 1999-2000 and 2000-01. This has affected overall gross state domestic product (GSDP) and per capita income which have declined during the year of 2000-01.

Table 2.10

Sectoral GSDP Over a Period of Time

(In ₹ Crores)

Year	*Disasters*	*Primary*	*Secondary*	*Tertiary*	*GSDP*	*Per Capita Income (₹)*
1993-94	Partial Drought	12525	17595	19074	49194	11323
1994-95		17047	20218	20793	58058	13104
1995-96	Partial Drought	15190	23255	22801	61246	13556
1996-97		19789	25954	24222	69966	15234
1997-98	Flood	18249	25962	27231	71442	15335
1998-99	Cyclone	19228	28119	29225	76571	16200
1999-00	Partial Drought	14210	30939	32205	77353	15785
2000-01	Drought	12693	30012	33241	75945	15157
2001-02	Flood	16238	30401	35628	82267	16047
2002-03	Flashflood	14436	35558	38430	88424	17040
2003-04		21695	38264	41835	101794	19265
2004-05		19756	41837	45440	107033	19899

Source: Centre for Monitoring Indian Economy (CMIE), November, 2006.

However, the economy of Gujarat could catch over the adverse impact of disaster and has shown an increasing trend in subsequent years. The external funding has played a significant role in eliminating adverse effects of disaster. However, it is very important to note here that such funding may not be available all the time, specifically when India has decided not to accept any international help after 2004 Tsunami.

Figure 2.3

Production of Primary Sector Over a Period of Time

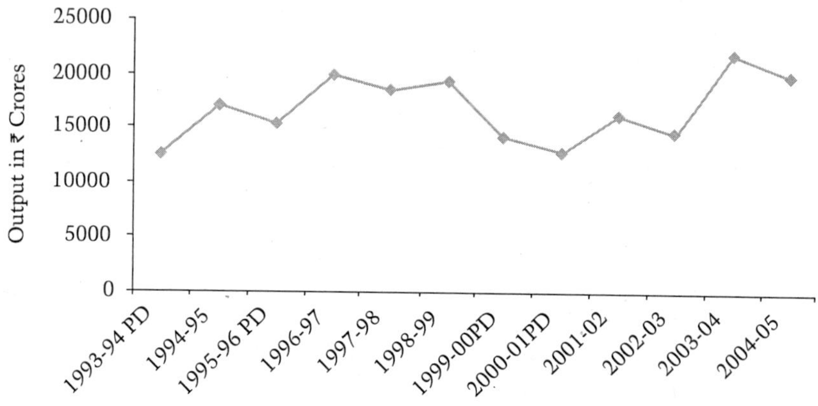

Note: PD is Partial Drought and FD is Full Drought.
Source: CMIE, November, 2006.

In short, the macroeconomic effects of disasters are strikingly visible in the state and the state has to rethink about its strategy for mitigating the potential risk due to future disasters.

The statistics in Table 2.11 very clearly show impact of drought on the state economy. Deficiency in rainfall in the years of 1995-96, 1999-2000, 2000-01 and 2002-03 has affected the food grain production adversely. Decline in food grain production has resulted in negative agricultural growth. However, such adverse effects are not shown at GSDP level mainly because the other two sectors have been doing better during these years.

Although disasters in the state have not affected an overall state economy very badly, there are some striking observations related to revenue losses. It is obvious that revenue generation declines during the time of disaster.

Table 2.11

Impact of Drought on Gross State Domestic Product in Gujarat (1994-95 to 2003-04)

Year	(% Deviation)				
	Rainfall decreased	Food Grain Area	Food Grain Production	Agricultural Growth	GSDP Growth
1994-1995	68.3	5.8	38.8	44.4	18
1995-1996	-54.1	-7.1	-17.1	-12.8	5.5
1996-1997	25.7	1	27.5	36.7	14.2
1997-1998	27.1	-0.7	0.4	-9.4	2.1
1998-1999	27.1	-4.3	-1.2	7.2	7.2
1999-2000	-42.8	-7.6	-26.5	-30.8	1
2000-2001	-18	-23.1	-42.6	-12.2	-1.8
2001-2002	71.6	12.8	92.5	31.5	7.7
2002-2003	-20.7	-7.4	-26.2	-15	10.1
2003-2004	60.4	25.5	81.5	77	15.4

Source: Directorate of Economics and Statistics, GoG.

In fact, the revenue losses had long-term impact over the years. Figure 2.4 indicates the loss of revenue receipt (RR) during 1999 to 2004-05. If the effects of drought in the years of 1999-2000, 2000-01 and 2002-03 as well

Figure 2.4

Growth Rate of Estimated Revenue Receipt and Actual Revenue Receipt in Gujarat

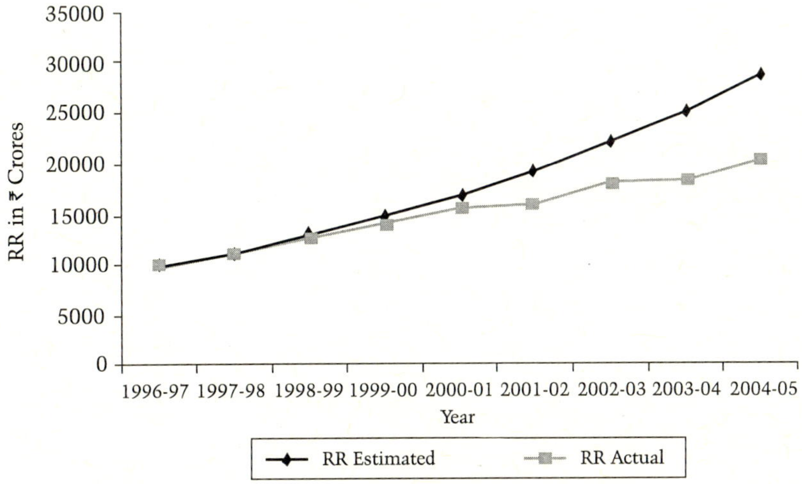

Source: Monthly Review of Gujarat Economy CMIE, November 2006.

as flood in the years of 2001-02, 2003-04 and 2004-05 are eliminated, the growth of revenue receipt might have been different than the actual. The tow lines show the estimated RR and the actual RR. It is also important to note here that the two consequent years of disasters can have substantial adverse macroeconomic effects.

Although the state has progressed significantly after these disasters, the macroeconomic impact of disaster on the state is starkly visible. The growth story of Gujarat could have been different if the impact of disasters is not felt by the state.

The above discussion is just based on very superficial observations. It is required to have an in-depth study to understand the macroeconomic impact of climate-induced disasters in Gujarat.

Intensity Indicators of Climatic Disasters

Three indicators of impact are derived in order to measure the risk intensity of climatic disasters. It is assumed here that the higher impact of disaster is mainly due to greater intensity. Table 2.12 denotes the coefficients of vulnerability. Coefficient of mortality is the highest for flood, which is followed by cyclone and drought. Paradoxical to this, coefficient of social vulnerability is the highest for drought and the lowest for cyclone. This shows that the social impact of drought is prevalent in the state.

Table 2.12

Indicators of Risk Intensity Due to Disasters

Risk Intensity Indicators	*Flood*	*Cyclone*	*Drought/ Heat Wave*
Average Death toll (M)	13463	5069	685
Coefficient of mortality (CM)	26.6	10	1.4
Number of people affected (SV)	256118523	10733839	391175000
Coefficient of Social Vulnerability (CSV)	5.1	0.2	7.7
Average Economic loss Eh in US$ (EV)	1901669	4323056	676359
Coefficient of economic vulnerability (CEV)	12.2	27.7	4.3

Source: Calculated by the author from EM-DAT: The OFDA/CRED International Disaster Database.

Majority of the rural population in the state is dependent on rainfed agriculture for their livelihood maintenance. The drought affects the livelihood of people to a great extent. It is interesting to note here that the coefficient of economic vulnerability is the highest for the cyclone. The coastal area of the state is highly prone to cyclone. Further, the port-led economy has developed the coastal area with higher economic growth. Densely populated and economically well-developed coastal region put the region more exposed to cyclones and therefore, the probability of economic damage due to cyclone is exponentially high in the state. In short, given a similar external shocks in terms of various climatic hazards, the risk varies greatly over space. This is mainly because the difference within the vulnerability aspects over space, community and regions.

Expenditure on Disasters

Although the state is responsible for declaring disaster in the state, it has to depend on the central government for funds and grants. Indian government has seriously looked into the adverse impact of disasters and has considered always separate fund for relief activities. The first attempt to keep separate fund for relief activities was the introduction of 'margin money scheme' by the Second Finance Commission while assessing the needs of the revenue for each state. The intention was to keep this margin fund (amount ranging between ₹ 10 and 100 lakh) for meeting the expenditure on natural calamities which mainly considers immediate relief measures. It was the Sixth Finance Commission which recommended for systematic development of the drought- and the flood-prone areas through plan programmes along with the continuation of the margin money arrangements. Thereafter, the central government financially assisted the state governments in the form of various types of contribution including margin money, advance plan assistance (grants and loans) and special assistance as grants and loans. In this way, the size of the margin money provided for the states increased gradually from ₹ 6.15 crore per annum (Second Finance Commission) to ₹ 240.75 crore per annum (Eighth Finance Commission).

The Ninth Finance Commission replaced the margin money scheme into Calamity Relief Fund (CRF) for each state. The fund was raised from the central and state governments on a sharing basis of 75:25 ratios. The size of the fund was determined by average past year spending on disaster relief in a particular state. This finance commission examined the importance of national insurance fund but did not recommend by giving a simple explanation that poor in our nation do not have any assets to insure.

Further, the concept did not receive much attention, as the commission observed that the concept was not viable or practicable on the ground that the process of making assessment of loss by an external agency bound to be complicated and time consuming. This would defeat the very purpose that is providing timely aid to the affected people.

The Tenth Finance commission observed that the fund created under CRF was not able to meet the losses created by severe disasters. Therefore, it recommended the setting up of a National Fund for Calamity Relief (NFCR) for helping the state in case of rare but severe calamities. Despite the two big fund-raising schemes, the Eleventh Finance Commission noticed that the fund under NFCR was used up just in three years and could not meet the expenditure required for the relief measures after big and rare calamities. Therefore it suggested for setting up of a National Calamity Contingency Fund (NCCF) from the extra levies of special surcharges of central taxes. It does not require the parliament consent as it falls under public account system.

The Twelfth Finance Commission recommended to extend the definition of calamities to cover landslides, avalanches, cloud burst and pest attacks. This commission also did not see the insurance scheme very viable, as it believed that the whole burden of insurance premium on the centre and state governments would put heavy burden. However, it recommended that the centre and state governments can play a vital role in encouraging insurance of private assets in vulnerable zones.

The latest and Thirtieth Finance Commission has further refined the financing of disaster relief fund and has recommended converting the NCCF into National Disaster Relief Fund (NDRF) which has to go to Parliament for due appropriation. Similarly, CRF fund needs to be merged into State Disaster Relief Fund (SDRF).

In short, the country has quite a good financing management for relief measures. However, utilisation of such funds does not include the long-term reconstruction as well as mitigation plans. For mitigation and risk reduction strategies-finance commission recommends states to use their plan expenditure under various departments.

The other important development in the field of disaster management happened when the Eleventh Five Year Plan document accentuated for mainstreaming disaster management into development planning. The report pointed out that development plans under all the departments have to ensure that they do not increase the potential risk and vulnerability towards future disasters. Although the country has well plan for disaster relief expenditure, it has lacked the financial strategies for disaster risk reduction and mitigation. Furthermore, there is no state-level initiation for developing disaster risk reduction strategies in India.

Figure 2.5 shows the expenditure of national and state budget on various calamities. The percentage share of calamities fund to total budget is relatively higher in Gujarat when compared to that of India. This indicates that the state is highly prone to natural disasters and has to spend more on relief and rehabilitation work. This is showing the serious problem of disaster in Gujarat and one has to think about risk management and risk transfer or risk reduction strategies.

Further, the state's share in national calamities fund which is used for relief expenditure is very high. During the earthquake time, the percentage share of the state became 50 per cent. However, gradually it declined and reached to 4 to 7 per cent only. This also confirms the severity of disasters in Gujarat.

Figure 2.5

Expenditure on National Calamities (% to Total Budget)

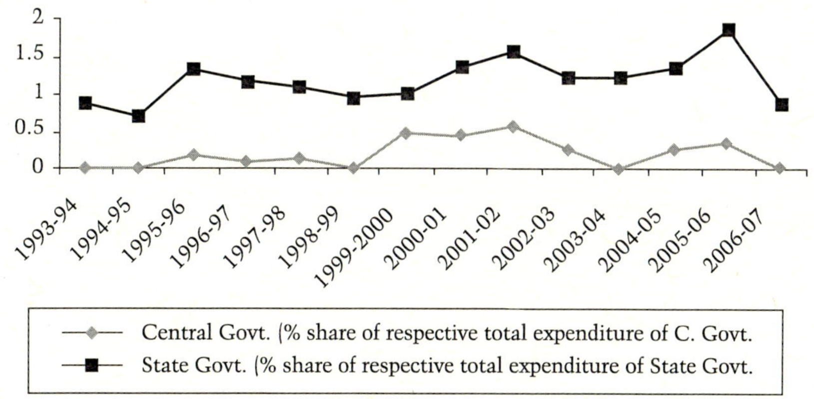

Source: CMIE, Public Finance.

Table 2.13

Expenditure on Disasters

All States	*Expenditure on National Calamities (in ₹ crore)*		
	Gujarat	*All States*	*Gujarat Share in Relief Expenditure*
1990-00	438	2612	16.8
2000-01	1476	3878	38.1
2001-02	2504	5012	50.0
2002-03	319	4169	7.7
2003-04	350	4542	7.7
2004-05	255	5390	4.7
2005-06	687	8810	7.8
2006-07	81	4516	1.8

Source: Revenue Department, GoG.

Figure 2.6

Expenditure on Disasters

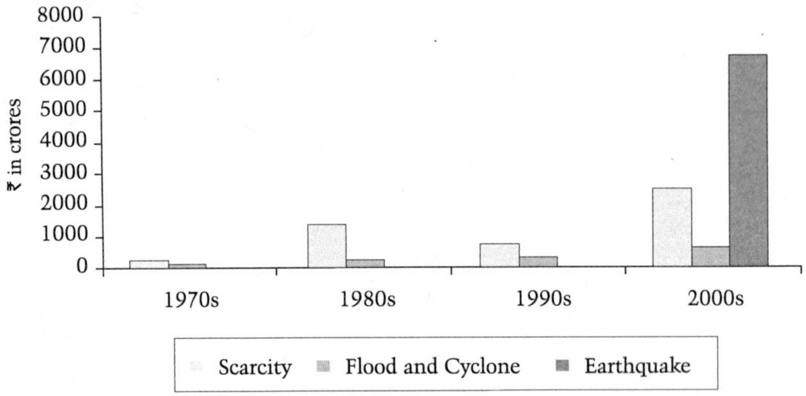

Source: Revenue Department, GoG.

Although Gujarat State Disaster Management Agency (GSDMA) is the nodal agency for managing disasters, its main focus is on earthquake rehabilitation project. Therefore, the data on expenditure made on various disasters in the last four decades are obtained by Revenue Department. Figure 2.6 presents the expenditure scenario of various disasters. Though the expenditure on earthquake is only one-time expenditure, it has exceeded to the total expenditure made on drought, flood and cyclones during the last four decades. This once again confirms the fact that though the state was disaster prone to various disasters, only earthquake hit during the year of 2001 made its vulnerability more visible. Flood and cyclone being major frequent and dangerous disasters are also important with regard to expenditure made on disasters.

Disaster Management in Gujarat

The declaration of disaster is a subject of state. The state revenue department declares the disaster at any level in the state on the recommendation of state relief commissioner or district collector. Hence, the Revenue Department is the nodal agency to declare disaster in the state. However, there are number of departments involved in dealing with disaster. The state has established a special autonomous department called Gujarat State Disaster Management Authority to deal with the disasters in a continuous and integrated manner. The authority prepared a disaster management policy which subsequently became a disaster management act in 2003. The act specifies the rules and responsibilities of various functionaries in the state to work during the period of disasters. It has laid down the guidelines for the state in which different departments have to incorporate disaster management plan under the supervision of State Disaster Management Authority (SDMA). In fact, the act specified the function of GSDMA which are the following:

1. To promote an integrated and coordinated system of disaster management.
2. To collecting and storing data on all aspects of disasters.
3. To promote awareness and preparedness to deal with potential disasters.
4. To prepare disaster management plans at various levels and to facilitate in the activities of relief, rehabilitation and reconstruction.

Disaster management practices in the state follow both short-term and long-term strategies. The short-term strategies include relief and reconstruction activities which are carried out just after the disaster strikes. The commissionaire of relief in GoG is the nodal agency to provide relief for all types of disasters. Long-term strategies to mitigate the adverse effects of disasters are mainly looked after by GSDMA. However, GSDMA mainly concentrates on rapid onset disasters. So far, it has extensively worked on earthquake which is non-climatic disaster. However, recently it has started working on flood and cyclone management practices. GSDMA has established early warning system and prepared disaster preparedness plan at various levels (state, district, block and village). It has also prepared building codes and vulnerability atlas. Vulnerability maps shown on the website of GSDMA have measured mainly physical vulnerability.

GSDMA, a nodal agency to work for disaster management, has contributed significantly towards working for managing disasters in Gujarat. The state of Gujarat received a huge grant from the World Bank and Asian Development Bank after the earthquake hit the state. The agency does not distinguish between climatic and non-climatic disasters. However, it is useful to assess the pattern of expenditure as far as disaster risk reduction is concerned. Figure 2.7 indicates the pattern of expenditure carried our by GSDMA up to the year of 2010. The agency spent a total of 2419 crores of rupees on various strategies of disaster management. Out of this total fund, the agency spent 93 per cent of the budget in rehabilitation activities. About 5 per cent of budget is spent on capacity building of the community and only 2 per cent is spent on disaster risk reduction. GSDMA has successfully implemented the school safety programme for capacity building under IDRS (Integrated Disaster Reduction Strategies). It is worth noticing here that the state has to focus on health and education sectors while carrying out not only reconstruction and rehabilitation programmes but also during implementing long-term strategies on risk reduction. The state has accepted risk transfer strategies as an important one. Therefore, it has to work upon it both at policy as well as implementation levels. It is worth mentioning here the efforts carried out by GSDMA. It has made housing insurance mandatory with housing grants and loans. This means that the state government has considered all the management measures important for disaster risk reduction. It has wonderfully worked on preparedness measures. It has also plan for formulating strategies on risk prevention or risk transfer.

The state has yet to work on disaster risk reduction. It is, therefore, important to work on risk assessment techniques to mitigate the potential risk due to climate-induced disasters.

Figure 2.7

Expenditure Pattern on Disasters in Gujarat

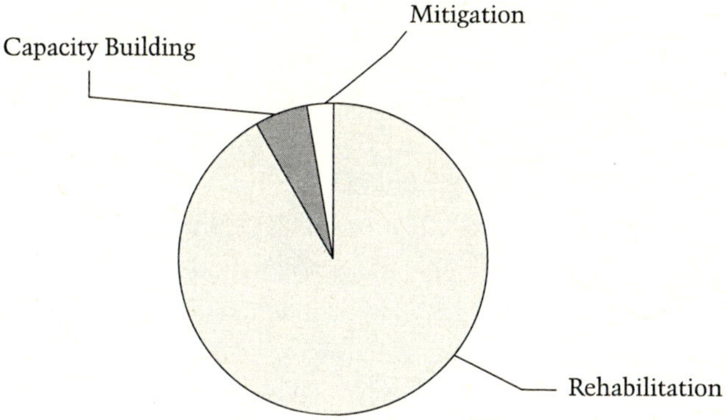

Source: Rajya Sabha Unstarred Question No. 4248, dated 30.4.2010.

There are other line departments involved in implementing long-term measures for disaster risk reduction. Long-term measures for drought mitigation are carried out by Rural Development Department. Some of the important mitigation strategies for drought include area development plans such as District Development Programme (DDP), Drought Prone Area Development Programme (DPAP), Watershed planning, natural resource management and employment guaranty programmes such as Mahatma Gandhi National Rural Employment Guarantee Act (MNREGA). The role of such schemes in mitigating drought impact has remained significant. With regard to flood mitigation activities, Flood Control Cell under Irrigation Department plays a major role. It identifies the area prone to flood for various reservoirs and dams in the state. It also outlines the guidelines for various functions for their roles and responsibilities during flood situation. In order to mitigate the risk of flood, the department has carried out various physical and structural mitigation works.

The implementation of such disaster risk reduction strategies for all these disasters heavily depends on area prone to disasters. Most of the disaster management plans prepared for particular disaster delineate the area prone to that disaster in the state. However, the risk and impact of disasters also depend on vulnerability pattern of the state. The process of assessing risk and implementing risk reduction strategies does not consider vulnerability issues while executing the risk reduction plan. Unless the vulnerability part is addressed properly, the risk reduction strategies would not remain successful. The risk calculated for each of these three disasters in this paper incorporates vulnerability measurement.

There are various non-governmental organisations (NGOs) working on disaster management programmes during the time of disasters. Some of them are worth mentioning here. OXFAM, CARE, ACTION AID and CRS have prepared an integrated programme for capacity building at organisation level. The United Nations Development Programme (UNDP) and other organisations have also prepared various programmes of disaster management. With respect to risk transfer strategies, various attempts have been made at small scale. The micro insurance implemented by Self Employed Women Association (SEWA) and All India Disaster Mitigation Institute (AIDMI) is worth noting in these cases. Banks like National Bank for Agriculture and Rural Development (NABARD) and Friends of Women's World Banking (FWWB) have also tried to provide micro insurance in the field of crop and livestock sectors. However, micro insurance provided by them is mandatory with micro credit or loan. Thus micro insurance basically works as loan or credit proofing rather than reducing vulnerability. There are other quite a few problems in implementing micro insurance at state level. These include fewer acceptances due to unawareness, moral hazards, absence of assured minimum insurance and problems of calculating premiums as well as damage assessment. In short, there are successful examples of micro-level insurance programmes run by BASIX, AIDMI and SEWA. The strength and constraints of such programmes should be assessed carefully. One has to proceed for scaling up such programmes by providing incentives and cross subsidies in order to reach the bottom layer of the poor.

Conclusion

The above analysis of effects of disasters on Gujarat state has clearly indicated that the risk due to climate-induced disasters is increasing over time not only in terms of frequency and intensity of the disaster but also in terms of its social and economic impacts. In other words, climate-induced disasters in Gujarat have put a lot of social and economic burden on the state economy. The financial implications of the disasters are prominently visible on the state economy.

Disaster management in Gujarat remained reactive by responding disasters through provision of relief during and after calamity. In other words, the state used to adopt a relief-driven disaster management system. However, in recent times, there has been a paradigm shift and the state has become or is becoming more proactive with emphasis on disaster prevention, mitigation and preparedness. The GSDMA has been set up as a nodal agency to plan for various strategies related to disasters. The GSDMA has not only worked extensively on relief and rehabilitation activities but has also focused on risk reduction strategies such as preparation of pre-disaster management plan at various levels. The state has shown a sincere commitment towards disaster management activities. Therefore, it is very important that the state looks into its financial planning for dealing with any future disaster. The state has to develop a concrete strategy for disaster risk reduction. *Ex-ante* measures such as traditional hazard insurance, parametric insurance, catastrophic pools, catastrophic insurance pools or government's natural catastrophic fund have to be considered for disaster risk reduction strategies. Such strategies need to be mainstreamed in development planning of the state.

3 Development and Disasters

A Status of Development in Gujarat

Disasters create a setback in the process of development. At the same time, the type of development can increase or reduce the risk due to disaster. Hence, disasters and development are mutually related. It is important to see that the pace of development does not increase the risk due to climate-induced disasters. It is, in this respect, essential to assess the present status of development in the state. This chapter assesses the existing status of development in the state of Gujarat. It has evaluated different dimensions of development. Various indicators have been analysed to assess the demographic, socioeconomic and environmental aspects of development.

Demographic Scenario

Demographic pressure can increase exposure of a region towards risk due to disasters. So it is important to assess the demographic trend and scenario in the state. Demographically, the state of Gujarat showed remarkable changes during the last decade i.e., 2001 to 2011. The decadal growth rate of population for the last decade in the state remained 19.17 per cent, which is higher than that of all India level (17.64%). Similarly the decadal growth rate in male and female population also remained high in the state when compared to respective figures at national level. In fact, the decadal growth rate of Gujarat during 1991 to 2001 remained at par with Maharashtra which is its neighbouring state. However, during the last decade, the decadal growth rate of population remained quite high (19.17%) when compared to that of Maharashtra which is 15.99 per cent. This means that state could not achieve much in terms of population stabilisation or one can say that the state has a long way to go to achieve demographic transition.

Table 3.1

Demographic Features of Gujarat and India

Demographic Features	*Gujarat* 2001	*India* 2001	*Gujarat* 2011	*India* 2011
Decadal Growth Rate of Population (Persons) (1991-2001, 2001-2011)	22.66	21.34	19.17	17.64
Decadal Growth Rate of Population (Male) (1991-2001, 2001-2011)	23.56	21.17	19.32	17.19
Decadal Growth Rate of Population (Female) (1991-2001, 2001-2011)	21.70	21.97	19.01	18.12
Density of Population (per sq. km.)	211	325	308	382
Level of Urbanisation	37.6	27.81	42.58	31.16
Literacy Rates (Persons)	69.14	64.8	79.31	74.04
Literacy Rates (Males)	79.66	75.3	87.23	82.14
Literacy Rates (Females)	57.80	53.7	70.30	65.46

Source: Census of India 1991, 2001 and 2011.

Compared to all India as well as its neighbouring states, the density of population is less in Gujarat. Nevertheless, the rate of increase in density over a period of time is quite high. Density of population in the state remained 308 in the year of 2011 which was only 136 in the year of 1971. This means that during the last four decades, the density of population became more than double.

Children are highly vulnerable to disasters as they are not able to protect themselves. The percentage share of (0-6) age group to the total population has declined from 2001-2011, its rate of decline is slightly less than that of all Indian level. The decline in the share of (0-6) age group of population for Gujarat is 2.46 per cent, while for all India level, it is 2.80 per cent.

The process of urbanisation increases the concentration of population at particular places. This accelerates the level of exposure towards disaster risk. Gujarat is considered as one of the highly urbanised states in India. The level of urbanisation has increased very fast in the state. The state reported about 42.58 per cent of people residing in urban areas as against 31.16 per cent of urban population at all India level. This shows that the state has shown a remarkable jump in proportion of urban population. The 2011 census has recorded a large number of new towns. The number of census towns has increased to 348 in 2011 from 242 in 2001. This is mainly because the 2011 census considered outgrowth as separate towns. This increased the number of towns drastically. In fact, the state experienced the high pace of urbanisation even in the last decade of the 20th century.

Table 3.2

Demographic Features of Districts in Gujarat

District	Percentage Decadal Growth Rate of Population	Sex ratio (Number of Females per 100 Males)	Child Sex ratio ([0-6]) Females per 1000 Males)	% of Urban Population	Literacy Rate
Ahmadabad	22.37	903	859	84.05	86.55
Amreli	8.59	964	879	25.49	74.49
Anand	12.57	921	877	30.32	85.79
Banaskantha	24.43	936	926	13.27	80.94
Bharuch	13.10	924	913	34.07	71.58
Bhavnagar	16.53	931	885	41.01	76.84
Dohad	29.95	986	937	8.99	50.60
Gandhinagar	12.15	920	847	43.21	85.78
Jamnagar	13.38	938	894	44.96	76.63
Junagadh	12.01	952	899	33.05	76.60
Kachchh	32.03	907	904	34.72	76.88
Kheda	12.81	937	887	22.72	84.31
Mahesana	9.91	925	845	25.35	84.26
Narmada	14.70	960	937	10.44	73.29
Navsari	8.24	961	91.4	30.74	83.03
PanchMahals	17.92	945	921	14.00	84.78
Patan	13.53	935	884	20.93	73.47
Porbandar	9.17	947	894	48.77	81.27
Raikot	19.87	924	854	58.12	82.20
Sabarkantha	16.55	950	898	14.96	74.40
Surat	42.10	788	836	79.68	85.55
Surendranagar	15.89	929	889	28.30	73.79
Tapi	12.07	1004	944	9.79	69.73
The Dangs	21.44	1007	963	10.89	76.80
Vadodara	14.16	934	890	49.54	66.39
Valsad	20.74	926	923	37.23	72.32

Source: Census of India, 2011.

The district-wise pattern of urbanisation shows that seven districts have shown higher percentage of urban population than the state average. Most of them are coastal districts lying in golden and silver corridor of the state. Tribal as well as north Gujarat districts have reported lower level of urbanisation in the state. Further, Rajkot, Bhavnagar, Vadodara, Bharuch and Valsad reported drastic increase in number of towns, while Ahmedabad has reported decline in both census towns and statutory towns. This is mainly because the urban agglomeration developed near Ahmedabad is merged into Ahmedabad city area.

The Status of Education

Literacy level can show the capability of community to cope up, adapt and reduce the risk. Highly literates are less vulnerable towards social and economic impact of disasters. Literacy level in the state is relatively better than that of all India level. The literacy rates in the state increased by 10.17 per cent from 2001 to 2011, while for all India level, it increased by 9.20 per cent. Further, the increase in literacy rates in female remained higher than in male for both the state and the country. This is mainly because the basic statistics of female literacy rates were quite low in the year of 2001. The state has to continue its efforts to achieve the percentage of literacy rates at par with other developed states such as Maharashtra and Andhra Pradesh.

Further, there is a wide variation in literacy rates within different regions of the state. Districts lying in the middle as well as south Gujarat regions have reported higher rates of literacy for both males and females. Tribal districts such as Dohad, Banaskantha, Tapi, Narmada and Panchmahals as well as drought-prone districts such as Kachchh, Surendranagar and Patan have shown less literacy levels compared to southern and middle part of Gujarat. In fact, some of the coastal districts such as Jamnagar, Amreli, Porbandar, Bhavnagar and Junagadh have reported literacy level below the state average. This indicates there is a wide regional variation in the status of literacy within the state. Vulnerability towards disaster and backward *talukas* are lagging behind in terms of achieving better literacy rates.

Education plays a crucial role in reducing vulnerability of people by enhancing personal endowment and capacity building to deal with risk due to disasters. It is also one of the three basic indicators showing the human development. Education enhances the potential capacity to capture new opportunities and also to maintain livelihood during the time of disaster. The government of Gujarat has remained highly successive in reducing dropout rates during the last decade. The dropout rates for class I to V declined from 21 per cent in the year of 2000-01 to 2.20 per cent in 2009-10. One of the crucial factors for this decline is mid-day-meal scheme which has been successful at least to attract the children in schools.

Table 3.3

Human Resource Availability for Education

	2000-01	*2009-10*
Pupil to Institution Ratio	222	204
Pupil to Teacher Ratio	45	35

Source: Census of India, 2001 and 2011.

Further, the state has developed its educational infrastructure and human resource base at par with number of children. This is clearly visible in the ratios of pupil to institutions and pupil to teachers for primary sections. Although the indicators showing the status of education are better, they do not represent the quality of education. Therefore, there are other outcomes indicators such as reading fluency, writing fluency and language competency that are important, as they can measure that if community can access the help provided to them before, during and after the disasters happen.

Inequality in Population

Inequality among men and women in a society spawns disproportionate status of women in terms of exposure, access to resources, capabilities and opportunities when compared to those of men (UNDP, 2011). Such lopsided status of women subsequently constructs vulnerability towards climate-induced disasters. Sex ratio is an important indicator of measurement showing an inequality between males and females in a society at a particular point of time. It denotes the ratio between males and females. In India, sex ratio is calculated as number of females for every thousand males.

Table 3.4

Sex Ratio in Gujarat and India

	Sex Ratio		*Juvenile Sex Ratio*	
	2001	*2011*	*2001*	*2011*
India	933	940	927	914
Gujarat	920	918	883	886

Source: Census of India, 2001 and 2011.

Lower the sex ratio, higher is the inequality between men and women in a society. The lower sex ratio indicates the preference towards boy child in a society. Moreover migration of male workers also reduces the sex ratio. Gujarat has shown very low sex ratios including the extremely lower juvenile sex ratio. This shows a clear bias towards male dominant society. Disasters intensify the inequality prevailing in a society. Women are more vulnerable because their multiple roles are directly affected at the time of disaster. As mentioned earlier, inequality in a society is an indication of male patriarchy, which excluded women from participating in decision making. In this way, resourceless and powerless women are highly vulnerable to risk due to disasters.

Looking into the data for the state and all India level, the sex ratios for Gujarat is lower compared to an average all India sex ratio. In fact, there is a slight decline in sex ratio from 2001 to 2011. This is in contrast with all India average situations where sex ratio has increased from 933 in 2001 to 940 in 2011.

Juvenile sex ratio is the sex ratio in the age group of 0-6. There is a marginal improvement in juvenile sex ratio in Gujarat during 2001 to 2011. However, juvenile sex ratio in Gujarat is 886, which is extremely low compared to all India level juvenile sex ratio and those of other neighbouring states.

District-wise variation in sex ratio reveals the fact that the developed districts such as Surat, Ahmedabad and Kachchh have reported the sex ratio lower than the state average. One of the reasons for this may be inflow of male workforce towards these highly industrial districts from other districts and other states. This is particularly true for Kachchh where juvenile sex ratio is better (913) but overall sex ratio is extremely less (907). The districts showing lower than state average juvenile sex ratio are Surat, Mehsana, Gandhinagar, Rajkot, Ahmedabad, Anand and Amreli.

Health Status in the State

A good health can help people to function well and children to grow and learn even during the time of crisis. Therefore, the health is considered to be the first line of defence at the time of disasters. Healthy society can absorb the shocks of disaster easily and can continue their lives after disasters happen.

Table 3.5

Health Statistics in Gujarat

Health		*2000-01*	*2009-10*
Birth Rate (SRS) (Calendar Yr.)	Per '000	25.2	22.3
Death Rate (SRS) (Calendar yr.)	Per '000	7.5	6.9
IMR (SRS) (Calendar Yr.)	Per '000	62	48
Primary Health Centres	Per 10,000 habitations	16	17
Community Health Centres	Per 10,000 habitations	4	4
Sub-centres	Per 10,000 habitations	113	112

Note: SRS, Sample Registration Sample; IMR, Infant Mortality Rates.
Source: *Socioeconomic Review, 2010-11.*

Health sector in Gujarat has not improved in line with its high economic growth. Gujarat has performed better than all India average statistics in terms of urbanisation, industrialisation and economic growth. The same has not happened in case of health sector. The birth rates and death rates are at par with all India level rates. The decline in birth rates and death rates in the last decade remained 2.9 and 0.6 per thousand people, respectively. IMR in the state has declined from 62 in 2000-01 to 48 (per 1000 live births) in the year of 2009-10. This decline is not better than all India level average of decline in IMR. The less achievement in health indicators is a clear implication of neglected sector in the state planning. This is observed in terms of the state contribution in developing health infrastructure. The number of primary health centres, community health centres and sub-centres per 10,000 habitations has not shown any change during the last decade. During 2000-01, the number of primary health centres per 10,000 habitations remained 16 which marginally increased to 17 during the year of 2009-10. Similar is the case of community health centres which reported only 4 per 10,000 habitations during the last whole decade. The number of sub-centres, in fact, marginally declined from 113 (per 10,000 habitations) in 2000-01 to 112 in the year of 2009-10. The state has to prioritise the health sector for setting up a healthy community which is less vulnerable to climate-induced disasters.

Economic Growth and Development in Gujarat

The state has shown a remarkable achievement in terms of economic growth. Per capita income in the state has increased almost three times in the last decade. Per capita income at constant price increased from 17,227 in 2000-01 to 49,060 in 2009-10. The composition of growth pattern shows that the state has experienced a better composition of growth for all the three sectors. Production of total food grain remained high throughout the last decade. In fact, the growth of cotton production has increased 6 times from 12 lakh bales in 2000-01 to 74.00 lakh bales in 2009-10. The cotton is the major crop which contributed significantly in showing the high growth rates of agricultural production in Gujarat. Further, the production of all seeds has also increased from 17.38 lakh tonnes in 2000-01 to 47 lakh tonnes in 2007-08. However, the oilseeds in the state are heavily rainfed and there is a great fluctuation in yearly production of oilseeds during the last decades. The years 2000-01, 2002-03, 2004-05 and 2009-10 reported decline in oilseeds production. Climatic variability during this decade has played a major role in the production of oilseeds. Therefore, it is important to study the impact of climatic variability in the production of oilseeds in Gujarat.

Table 3.6

Socioeconomic Indicators of Development in Gujarat

No.	*Item*	*Unit*	*2000-01*	*2009-10*
1	Gross State Domestic Product (GSDP)			
	At current prices	₹ in Crore	111139	429356
	At constant prices (1999-2000)	₹ in Crore	104494	331633
2	Per Capita Income (NSDP)			
	At current prices	₹ in Crore	18392	63961
	At constant prices (1999-2000)	₹ in Crore	17227	49030
3	Agriculture Production			
	Total Foodgrains	Lakh Tonnes	31.85	56.05
	Cotton	No. of Lakh Bales	12.83	74.01
	Total Oilseeds	Lakh Tonnes	17.38	30.1
4	Livestock Production			
	Milk	Lakh Tonnes	53.17	88.43
	Eggs	Nos. in Lakh	3460	127.62
	Wool	Lakh kg	27.4	29.19
5	Annual Survey of Industries			
	No. of Factories	Nos.	14090	NA
	Employment	Nos.	752013	NA
	Net value added	₹ in crore	16856	NA
6	Factories (Calender Year)			
	No. of working factories (P)	Nos.	20424	24453
	Employment daily therein (P)	Nos.	866720	1257957

contd...

...*contd*...

7	Electricity			
	Installed Capacity	MW	8582	11758
	Generation	MUs	50506	69883
	Consumption	MUs	34325	56973
	Per capita Consumption	Unit	953	1491
8	Ports - Cargo Handled by			
	Major Port (Kandla)	Lakh Tonnes	367.41	795
	Minor & Intermediate Ports	Lakh Tonnes	731.8	2055
9	Education			
	Dropout Rate			
	I to V	%	20.93	2.20
	I to VII	%	38.92	8.66
10	Primary Education			
	No. of Institutions	Nos.	36830	42145
	No. of Teachers	Nos.	195919	244331
	No. of Pupils	000	8186	8601
	Pupils/Teacher Ratio	Nos.	42	35
11	Secondary & Higher Sec. Edu.			
	No. of Institutions	Nos.	6343	9299
	No. of Teachers	Nos.	64076	86775
	No. of Pupils	000	2247	3045
	Pupil/Teacher Ration	Nos.	35	35
12	Higher Education			
	No. of Institutions	Nos.	556	1405
	No. of Teachers	Nos.	11041	20054
	No. of Pupils	000	419	643
	Pupil/Teacher Ration	Nos.	38	32
13	Health			
	Birth Rate (SRS) (Calendar Yr.)	Per'000	25.2	22.3
	Death Rate (SRS) (Calendar yr.)	Per'000	7.5	6.9
	IMR (SRS) (Calendar Yr.)	Per'000	62	48
	Primary Health Centres	Unit Nos.	1001	1105
	Community Health Centres	Unit Nos.	244	291
	Sub-centres	Unit Nos.	7274	7274
14	Employment			
	No. of Employment Exchange	Nos.	37	44
	No. of Vacancies Notified	In'000'	113	211
	Placement in Employment	In'000'	80	178
	No. of Job Seekers at the end of December	Nos.	1067751	895936
	Total Employment	In Lakhs	16.22	19.82
	Private Sectors	In Lakhs	7.42	11.96
	Public Sectors	In Lakhs	8.80	7.86

contd...

...contd...

15	Public Finance			
	Capital Receipts	₹ in Crore	13904.38	14531.66
	Revenue Receipts	₹ in Crore	15738.59	41672.37
	Total Receipts	₹ in Crore	29642.97	56204.02
	Capital Expenditure	₹ in Crore	9103.42	11719.41
	Revenue Expenditure	₹ in Crore	22040.82	48638.27
	Total Expenditure	₹ in Crore	31144.24	60357.68
	Development Expenditure	₹ in Crore	19610.98	40418.4
	Non-Development Expenditure	₹ in Crore	11483.62	19833.31
	Revenue Deficit (-)/Surplus(+)	₹ in Crore	(-)6302.23	(-)6965.90
	Fiscal Deficit	₹ in Crore	7965.27	15153.27
	Total Tax Receipts	₹ in Crore	10620.57	32631.16
	Share of Central Taxes	₹ in Crore	1573.75	5890.90
	Sales Tax/VAT	₹ in Crore	5942.73	18199.79

Source: *Socioeconomic Review of Gujarat, 2010-11.*

As far as livestock production is concerned, the productions of milk and eggs have shown a remarkable improvement since 2000-01. The production of milk increased gradually from 53.17 lakh tonnes in 2000-01 to 88.53 lakh tonnes in 2009-10. The increase in milk production is continuous and gradual showing an improvement in dairy industry. In fact, the production of eggs has also increased to 12762 lakh in 2009-10 from 3460 lakhs in 2000-01. This indicates that the climatic variability does not have much impact on dairy industry. An overall good rainfall has contributed to this remarkable increase in livestock production. The livestock sector is less vulnerable towards climatic variability.

Table 3.7

Agricultural and Livestock Production in Gujarat

	Unit	*Year*			
		2000-01	*2001-02*	*2002-03*	*2003-04*
Agriculture Production					
Total Food Grains	Lakh Tonnes	31.85	52.54	43.95	67.37
Cotton	No. of Lakh Bales	12.83	16.85	18.84	42.80
Total Oilseeds	Lakh Tonnes	17.38	37.47	18.78	58.50
Livestock Production					
Milk	Lakh Tonnes	53.17	58.76	60.89	64.21
Eggs	Nos. in Lakh	3460	3701	3848	4423
Wool	Lakh Kg.	27.40	28.08	27.11	27.80

Source: Socioeconomic Review of Gujarat, 2010-11.

The share of industrial sector to the growth of economy is a substantial one. The employment in industrial sector has increased to 11.25 lakhs in 2009-10 from 7.5 lakhs in 2000-01. Further, value addition from the industrial sector has also increased from ₹ 16856 crores to ₹ 61025 crores during the last decade. The higher level of per capita income also leads to an increase in use of power consumption. The per capita consumption of power increased from 953 units to 1491 units in this last decade.

Briefly, the state has not faced big disaster in the last decade; the economic growth has not been affected at macro level. However, climatic variability has a significant impact on crop production which can affect the growth of economy of the state.

Similarly, the development of ports has significantly contributed towards higher income growth. It seems that the minor and intermediate ports have contributed significantly by reporting three times increase in cargo handled by these ports during the last decade. It is interesting to note here that during 1990s, the role of major ports was substantial in the growth of the state economy (Lodhia, 2010). There is devolution of development due to pro-industrial strategies in the coastal areas of Gujarat.

Status of Natural Resources and Environment

The state has reported a significant change in its stock of natural resources over a period of time. This section has assessed the status of natural resources and environment in terms of quantity and quality. Three basic natural resources, namely land, water and forest, are evaluated in terms of change in quality and quantity over a period of time. Table 3.8 depicts the change in land use in the state during 1980 to 2005.

The table shows that the area under forest has continuously declined over a period of two and a half decades. This decline is more visible in the decade of 1980s. The state has acknowledged the conservation of forest in various development programmes. Yet, the state has not been able to recover the lost forest areas. Another important change is noticed in an increase in barren and uncultivated land. As far as primary production is concerned, area under food crops has declined significantly. This is mainly because farmers are turning towards non-food crops which are less volatile towards climatic variability. Further, area under current fallow and area under non-

Table 3.8

Change in Land Use in Gujarat

Land Category	Percentage Change Over a Period of Time				
	1980-90	1990-95	1995-2000	2001-05	1985-05
Forests	-4.1	0.0	0.1	-0.6	-5.7
Barren and Uncultivable Land	4.2	0.0	0.0	0.3	4.2
Area under Non-Agricultural Uses	5.2	0.0	0.4	0.4	7.5
Cultivable Waste	-0.8	0.0	0.7	-0.4	-0.5
Permanent Pasture and other Grazing Land	-0.3	0.0	0.3	0.0	0.3
Land under Miscellaneous Tree Crops and Other Groves not included in Net Area Sown	-2.4	0.0	0.0	0.0	-2.4
Current Fallows	92.4	-0.3	22.8	-27.1	24.2
Other Fallows	-84.3	-0.5	-52.4	-6.1	-96.3
Net Area Sown	-2.9	0.0	-2.4	3.3	1.8
Area under Food Crops	-4.2	0.0	-14.5	10.9	-10.8
Area under Non-food Crops	2.3	0.1	4.3	4.6	21.0
Gross Cropped Area	-1.0	0.0	-4.5	7.2	4.8
Total Reporting Area	0.0	0.0	0.0	0.0	-0.1

Source: Department of Agriculture, GoG.

agricultural uses have reported an increase since 1980. This shows the state is in the process of conversion of agricultural land into culturable wasteland. Pro-industrial policies and special economic zones have played a major role in acquiring good agricultural land for industrial uses. Further, gross cropped area has not shown any significant increase. In fact there was a decline in gross cropped area during 1995 to 2000. This was mainly due to two consecutive droughts faced by the state. The positive in area under food crops during 2001 to 2005 is the effect of less irrigated area in a base year because of drought. In short, the state has reported a remarkable change in land use categories. The change is mainly towards converting fertile agricultural land to culturable wasteland for industrial purpose. This, in long term, may not protect the livelihood of rural poor during the time of disaster.

The latest desertification and land degradation atlas of India has reported that about 68.43 per cent of geographical area is either degraded or in the process of desertification. The highest proportion of land (about

34.6%) is affected by water erosion. This is followed by salinity (16.8%) and vegetal degradation (14.0%). Porous geological soil along with over use of ground water has accelerated salinity ingression in the state. This has resulted in an increase in saline land in the state. Vegetal degradation is mainly observed in terms of deforestation, forest blanks, shifting cultivation and degradation in grass land and scrub land. Further, wind erosion has also degraded the land in Gujarat. The data show very clearly that the state is loosing its land quality due to various anthropological pressures. Degraded land resources force people and region towards the state of severe vulnerability during the time of disasters.

Table 3.9

Desertification or Land Degradation in Gujarat, 2010

	Area in Hectares	
Water Erosion	6790469	34.6
Vegetal Degradation	2737866	14.0
Wind Erosion	543322	2.8
Frost shat	0	0.0
Salinity	3294079	16.8
Mass Movement	0	0.0
Water Logging	0	0.0
Rocky and Barren	23521	0.1
Others	25951	0.1
Total	13415208	68.4
	19602400	100

Source: Desertification or land degradation atlas of India, SAC, 2010.

Similarly water resources in the state have also shown a remarkable change over a period of time. About half of the *talukas* in the state have reported their ground water development not in the safe category. As far as ground water development is concerned, around 14 per cent of *talukas* are categorised as over exploited. Table 3.10 represents the number of *talukas* falling under various stages of ground water development. Approximately 5 per cent *talukas* fall under the category of critical stage ground water development. Further, roughly 28 per cent *talukas* fall under the category of semi-critical stage of ground water development.

Table 3.10

Stages of Ground Water Development in Gujarat

Ground Water Development	*No. of Talukas*
Over-Exploited Category (>100% Ground Water Development)	30
Critical [Dark] Category (Between 90% and 100% Ground Water Development)	12
Semi-Critical [Grey] Category (Between 70% and 90% Ground Water Development)	63
Safe	117
Total	222

Source: Ground Water Resource Development Corporation, 2002.

In short, the state has faced challenges in maintaining its critical resources land and water both. It has to conserve these resources now as to provide livelihood to poor at the time of any external crisis, specifically at the time of climate-induced disasters.

Conclusion

The state has shown a remarkable economic growth in the last decade. Along with this growth, it has reported higher rates of growth of population and urbanisation. In terms of educational level, the state has achieved significantly in the last decade. However, health sector is still lagging behind. The most important challenge faced by the state is to conserve its natural resources. The quantity and quality of land and water resources are degrading. It requires an urgent attention to maintain these resources in order to sustain the development of the state in a longer run.

4 Determinants of Vulnerability in the State

A Macro Level Model

As mentioned earlier in this study, extreme climatic variability is one of the major implications of climate change. Such variability is often reflected in climate-induced disasters. The potential risk due to such climate-induced disasters also depends upon the existing pattern of vulnerability in the state. In this context, measuring vulnerability due to climate-induced disasters is indispensable for assessing risk and deliberating risk reduction strategies. Assessing vulnerability and risk due to climatic disasters requires a methodological approach. A promising *modus operandi* for assessing vulnerability lies in recognising the determinants of vulnerability prevailing in the state. It can expose many adverse effects which are led by climatic variability on society. Therefore, it helps state as well as society for preparing and adopting adequate and successful measures for reducing the risk factors. In other words, quantifying vulnerability can show direction for preventing losses and adaptation towards climate change in a long way. Therefore, measuring vulnerability can help decision makers in prioritising their intervention for risk reduction and climate change adaptation strategies under development policies. This can facilitate mainstreaming disaster risk reduction into development goals.

This chapter has attempted to identify the vulnerability indicators at state level. It further goes beyond identifying the indicators of vulnerability and also endeavours to quantify those indicators at district level. It then explores a regional model for assessing the pattern of vulnerability within the state and identifying the determinants of vulnerability existing in the state. Later on, the vulnerability assessment is applied to measure the risk due to climate-induced disasters. Such an exercise is useful for identifying the most vulnerable district in the state for regional planning. Although indicators are not comprehensive like the concept, yet they tend to help policy makers by deriving relative numbers and helping in prioritising the planning process in a relative ranking.

Climatic variability in this study is measured in terms of annual variability in weather for a longer period of time. Rainfall, being an important climatic characteristic, is used here as an indicator of climate. If the variability in annual rainfall reaches an extremity point, it is experienced in terms of climatic disaster. In other words, the climatic disasters are characterised by extreme variability in rainfall for a longer period of time. Therefore, the climatic extremity is observed as annual variability in rain fall, which is an important climatic characteristic.

Disasters can be classified in various ways. Chapter 2 has appraised various classifications as per their importance. Classification of disasters also depends on number and type of disasters happening on different geographical setup. A way of classification varies according to different levels of spatial units of analysis. At global level, one requires to have an elaborate classification. National level classification depends on an extent of disasters happening in various states.

Although all the three disasters selected in this study are influenced by a common factor which is climatic variability, they can further be distinguished by the nature of an event. Hazards those which strike suddenly and cannot be predicted well in advance convert into rapid onset disasters. In this study cyclone and flood can be categorised as occurrence of rapid onset climate-induced disasters. It is impossible to forecast such events well in advance. Forecasting time varies from few hours to few days. On the other hand, slow onset disasters can be predicted well in advance. In fact, they are the implications of continuous occurrence of natural events resulting in disasters. Drought can be categorised under slow onset disasters. Planning commission defines drought-prone area as per the pattern and quantum of rain/precipitation. It declares drought when the rain during the main crop season of an area makes traditional cultivation of main crops hazardous in three years or more out of every 10 years (GoI, 1981). It implies that the drought declaration is the process of long span scarcity of water in a region.

All the three selected disasters are different as far as short-term impact is concerned. Deaths due to cyclone and flood are common and happen immediately, unlike drought. However, the consequent events of drought can lead to food insecurity for marginalised people. In addition to this, an extreme poverty and mal nutrition can sometimes lead to the death-related consequences. In spite of having their distinct nature, the risk due to these disasters is still a function of intensity of hazardous occurrence and vulnerability of the region. Therefore, the process of vulnerability assessment remains same for all the three selected disasters.

Risk Assessment for Gujarat: A Methodological Approach

It is a fact that the risk due to climate-induced disaster depends on probability and intensity of extreme events and prevailing vulnerability in a region. Probability and intensity calculation at state level has been carried out in previous chapters. As there is no systematic data available on climatic hazard at district level, the study has used alternative approach. While dealing with disasters, the state looks into an area prone to each disaster separately. In fact, each one of these disaster is dealt by different departments. Methodology for delineating an area prone to each disaster is separate and carried out by concerned departments. An area prone to each disaster has been delineated mainly based on probable hazard risk in different districts. Such methodology is explained in this chapter while calculating risk separately for each disaster.

Further, risk also depends on the vulnerability of the region which is exposed to the selected climatic disasters. Therefore, vulnerability index is carried out separately. Vulnerability by its simple definition refers to the susceptibility of a person, group, society, region or system to any external risk. In this context, external risk is defined in terms of climatic variability. Such variability is manifested by the climate extremes in terms of climatic disasters. In fact, it is not possible to define vulnerability precisely. The concept of vulnerability provides flexibility to accommodate a number of phenomena under it. Therefore, the term vulnerability is relative and explanation of the term depends upon various factors. The United Nations International Strategy for Disaster Reduction (UNISRD, 2004) defines the vulnerability as "the condition determined by physical, social, economic and environmental factors, which increase the susceptibility of a community to the impact of hazards".

Vulnerability indicators need to be robust, evident, self-expressive, replicable, comparable and easy to understand. The selection of indicators is a technical challenge. Based on United Nations Development Programme's (UNDP's) approach and other literature, the indicators are selected. For instance, the Human Development Index (HDI) is an important indicator of vulnerability which explains the status of human condition in a given region. This exercise is limited to the indicators which could be derived based on availability of information at district level in the state of Gujarat. Some of the indicators are inversely related to vulnerability. For instance, an area under forest can protect the region and community from an intensive effect of climatic disasters.

In order to derive vulnerability index in this study, various demographic, physical, economic, social and environmental indicators have been identified and clubbed together. The indicators selected under each category of factors are shown in Table 4.1. About 42 indicators have been selected under four major groups which are responsible for existing vulnerability in the state of Gujarat.

The study has collected district-wise data of all the significant socioeconomic and environmental parameters. Such data on elements at risk are collected from various concerned state level departments or from published sources. Table 4.1 represents the selected indicators along with basic statistics. It shows that the coefficient of variation is extremely high for the indicators such as level of urbanisation and industrialisation. This indicates that the regional disparity is very high in the state in case of urbanisation and industrialisation. Further, a forest cover is unevenly distributed in the state making regional variation very high. On the other hand, some of the variables have indicated low disparity within the state. Such variables are female literacy rates, juvenile sex ratio and villages having no high school. The status of all these three indicators is poor in the state. As far as environmental indicators are concerned, the disparity is quite high making some district more vulnerable. High disparity in many selected variables among the districts shows the significant regional variation in the status of vulnerability. This signifies the fact that it is worth assessing the vulnerability status within the state and incorporating such measurement into risk calculation.

As the number of indicators is large and overlapping, the principle component analysis was carried out to reduce the number of indicators and derive the determinants of vulnerability. SPSS (Statistical Package for Social Science) was used for carrying out principle component analysis method. The major aim to use this technique was to reduce the complexity of information and reduce the observed variables by clubbing them together in a hidden factor which represents the number of variables. The variables were scaled up in order to compare with each other. The formula used for scaling up remained X = X – (Mean-2SD)/SD for each data set.

Thereafter, vulnerability index has been calculated by giving weightage equivalent to eigen value to each determinant of vulnerability. The model derived four major regional determinants of vulnerability existing in the state. These are vulnerable people, vulnerable surrounding, exposure to climatic hazards and vulnerable resources. These are explained in detail in the next section.

Table 4.1

List of Indicators Derived for Quantifying Vulnerability

No.	*District*	*Mean*	*SD*	*CV*
	Demographic Indicators			
1	Density 2001	345	187	54.3
2	Decadal GR 1991-2001	20.8	8.1	38.8
3	GR of Urban Population	21.6	29.9	138.3
4	Dependency Ratio	1.0	0.2	17.5
5	Pace of Urban Growth	47.5	74.4	156.5
	Economic Indicators			
6	Poverty	41.8	17.5	41.8
7	% of Agri Labour	25.7	6.8	26.4
8	Per Capita ML Fixed Investment	14057	23769	169.1
9	Per Capita SS Fixed Investment	1223	897	73.3
10	% of Nonirrigated area	68.8	16.8	24.4
11	Per capita bank deposit, 2000	7495	5922	79.0
12	Credit (₹ per capita), 2001-02	3837	4478	116.7
13	SS Unit per 1000 Population	3	2	64.6
14	ML Scale Unit per 100000 Population	3	3	110.9
	Social Indicators			
15	Percentage of SC/ST population	6.7	3.2	47.7
16	Adult Literacy Rate	66.0	9.1	13.8
17	% of Female Illiteracy	80.5	5.1	6.3

contd...

...contd...

18	Net Enrolment Ratio	66.7	12.8	19.1
19	Juvenile Se Ratio	890.0	43.0	4.8
20	HDI	0.5	0.1	14.0
21	GDI	0.5	0.1	18.2
22	GEI	0.3	0.1	21.8
23	IMR	67.3	13.5	20.1
24	TFR	2.8	0.6	21.1
25	% Villages not having tap water	43.7	21.9	50.2
26	% Villages not having electricity	23.2	14.9	64.3
27	% Villages not having Sanitation facility	62.5	14.7	23.5
28	% villages not connected with pucca roads	8.3	4.4	53.3
29	% villages having no high school	95.2	3.6	3.7
30	% villages sub-centres do not have buildings	54.7	14.0	25.6
31	Road density (per sq. km)	0.54	0.16	30.3
32	% of households not having drinking water	77.3	10.7	13.9
33	% of households not having electricity	23.2	14.9	64.3
34	% of households not having toilets	62.5	14.7	23.5
35	% of households with Kachcha house	46.2	21.4	46.3
	Environmental Indicators			
36	% of degraded land	13.0	8.3	63.9
37	% share forest area to geog. Area 1997	12.2	16.8	137.5
38	Level of ground water Development% 1997	75.4	37.4	49.6
39	% of villages having excess Fluoride	14.3	13.5	94.9
40	% of villages having excess Nitrate	3.7	3.4	92.6
41	% of villages having excess Salinity	7.1	6.5	90.5
42	inter district variation in relative index o f development	0.3	0.1	47.7

Note: SD- standard deviation, CV – coefficient of variation, GR - growth rate, ML - medium and large, SS - small-scale, SC/ST - scheduled castes and scheduled tribes, GDI - Gender Development Index, GEI - Gender Empowerment Index, IMR, infant mortality rates, TFR - total fertility rate.

Sources: Census of India, 2001, *District Statistical Outline*, 2001.

Determinants of Vulnerability

As discussed in the earlier section, vulnerability indicators are identified and measured through principle component analysis. Indicators are operational representation of attributes showing qualities or characteristics of vulnerability. Identification of indicators is carried out by refereeing various studies which showed a contribution of inherent characteristic of a society in increasing the external risk. Such characteristic includes various dimensions of a society or a region including socioeconomic, environmental, political and institutional aspects. The first step towards analysis was to club overlapping indicators and identify the drivers of vulnerability pattern in Gujarat. The analysis resulted into four determinants of vulnerability which persuaded the vulnerability pattern in the state of Gujarat. These four determinants are vulnerable people, vulnerable surrounding, exposure and vulnerable resources. Figure 4.1 represents the conceptual model depicting determinants of vulnerability towards climatic variability. Let us examine how each determinant of vulnerability derived from an exercise is contributing towards an increased risk of climatic disasters.

Figure 4.1

Measuring Vulnerability: A Conceptual Model

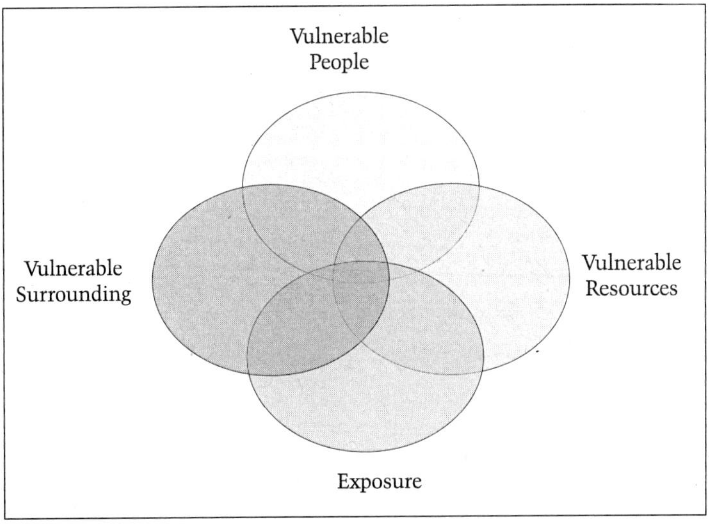

1. *Vulnerable people*: The first important determinant of vulnerability derived from the exercise is vulnerable people. Almost 35 per cent of variation in variables is explained by this determinant. The variables under this determinant mainly represent the marginalised groups of the society. These are proportion of dependents, proportion of schedule caste and schedule tribe, persons below poverty line, female agricultural labour, destitute and illiterate females. These variables capture an inability to recover from or absorb the shock of climatic disasters. Higher numbers in these

variables make the district more vulnerable. Although the climatic risk remains same, such identified groups would face higher risk than the other elite groups of the society. In other words, the impact of climate variability will be disproportionate on vulnerable people. Vulnerable people are mainly marginalised groups with relative inherent socioeconomic weaknesses which increase the direct impact of climatic disasters. Some of the other indicators falling into this category are outcome indicators which include HDI, Gender Development Index, gender empowerment measure and infant mortality rates. Apart from gender empowerment measure, all the other three indicators are negatively related. Lower values of these indices make the district more vulnerable. These indicators are characterised mainly by human development approach, which defines development as a process of enlarging people's choices. It puts people at the centre of development. These indicators represent capabilities of the community to adapt to new conditions. Lower indices of human and gender development represent less capability and fewer choices for people to cope up with climatic variability. In this way, vulnerability is determined by the level of human development of the community. The vulnerability explained by these variables has a major role in determining vulnerability pattern of the state. These variables have played a significant role in making backward districts such as Dahod and Panchmahal and drought-prone districts such as Kachchh and Surendranagar highly vulnerable.

2. *Exposure:* The second set of determinants of vulnerability is mainly relating to an exposure which incorporates the growth of population and economy. About 13 per cent of variation in variables is explained by this determinant. It is obvious that the densely populated areas with high industrial growth and even higher economic growth are more vulnerable. These indicators portend the region's susceptibility to climatic disasters whatever their type or intensity is. Most of the districts having higher exposure are located in the coastal areas, for instance, Jamnagar, Rajkot, Porbandar and Amreli.

Table 4.2

Ranking of Districts by Major Drivers of Vulnerability

Sr. No.	District	Vulnerable people	District	Vulnerable surrounding	District	Exposure	District	Vulnerable resources
1	Dohad	26.5	Dohad	16.4	Gandhinagar	15.2	Sabarkantha	15.6
2	Dangs	24.8	Dangs	15.3	Ahmedabad	14.3	Valsad	15.6
3	Navsari	21.8	Navsari	13.9	Jamnagar	14.1	Vadodara	14.3
4	Panchmahals	21.1	Surendranagar	12.4	Rajkot	13.2	Junagadh	14.3
5	Surendranagar	15.0	Panchmahals	11.9	Patan	12.9	Panchmahals	13.3
6	Narmada	13.4	Kachchh	10.0	Porbandar	10.5	Patan	12.9
7	Kachchh	12.6	Jamnagar	9.6	Amreli	10.4	Ahmedabad	12.9
8	Vadodara	10.6	Valsad	9.6	Mehsana	10.2	Bharuch	12.7
9	Mehsana	9.1	Vadodara	8.8	Bhavnagar	8.5	Kachchh	11.5
10	Banaskantha	8.8	Narmada	8.6	Junagadh	8.0	Surat	11.0
11	Sabarkantha	8.5	Surat	8.4	Bharuch	7.9	Amreli	9.9
12	Jamnagar	8.0	Bharuch	8.2	Sabarkantha	7.7	Anand	8.9
13	Junagadh	8.0	Junagadh	8.2	Banaskantha	7.7	Kheda	8.9
14	Valsad	7.4	Bhavnagar	8.1	Anand	7.5	Porbandar	8.9
15	Amreli	7.1	Banaskantha	8.0	Narmada	7.3	Narmada	7.9
16	Surat	6.1	Mehsana	8.0	Vadodara	7.2	Gandhinagar	7.7
17	Gandhinagar	5.8	Porbandar	7.7	Navsari	7.0	Rajkot	6.9
18	Bhavnagar	5.2	Anand	6.9	Kheda	6.8	Dangs	6.4
19	Bharuch	4.1	Amreli	6.7	Kachchh	6.8	Navsari	6.4
20	Kheda	1.8	Sabarkantha	6.5	Panchmahals	6.7	Mehsana	6.0
21	Patan	1.4	Kheda	6.3	Dangs	6.5	Surendranagar	6.0
22	Anand	0.8	Rajkot	5.5	Valsad	6.3	Banaskantha	5.1
23	Porbandar	0.7	Gandhinagar	4.8	Surendranagar	5.5	Dohad	3.0
24	Ahmedabad	0.2	Patan	4.2	Surat	5.1	Bhavnagar	2.3
25	Rajkot	0.1	Ahmedabad	2.3	Dohad	3.1	Jamnagar	1.8

Source: Calculated by the author from various published sources, 2001.

3. *Vulnerable surrounding:* The third set of variables which shapes the vulnerability is called vulnerable surrounding. About 10 per cent of variation in variables is explained by vulnerable surrounding. An easy access to basic services for human being provides better surrounding. Lack of such services makes people of that region more vulnerable towards the selected disasters. This is because lacking accessibility creates an adverse situation for people at the time of climatic disasters. The important basic services in this

category involve access to drinking water and sanitation, access to health and education and access to *pakka* road and electricity. When disaster strikes, an inadequate access to such basic services not only increases the impact of disasters but also reduces the chances of survival. These variables have played a major role in determining vulnerability in backward districts such as Dohad, Dangs, Panchmahal and Surendranagar as well as coastal districts such as Kachchh Jamnagar and Valsad.

4. *Vulnerable resources:* The forth category derived from the analysis was named as vulnerable resources of the region which make it more vulnerable towards the selected disasters. About 7 per cent of variation in variables is explained by this determinant. The contributory indicators in this category of determinants are degraded forest land, waste land, overused ground water and deteriorated quality of drinking water sources. Degraded resources do not support the community livelihood at the time of disaster such as drought, flood and cyclone and therefore make people of that region more vulnerable towards disasters. In other words, deteriorated resources create an unfavourable situation during the time of climatic disaster. In Gujarat, the poor base of natural resources in Sabarkantha, Valsad, Vadodara, Junagadh and Panchmahal puts these districts in vulnerable region.

Table 4.3

Relative Vulnerability Index at District Level in Gujarat

District	*Vulnerability Index*	*District*	*Vulnerability Index*
Panchmahals	13.3	Mehsana	8.3
Dangs	13.2	Bharuch	8.2
Dohad	12.3	Patan	7.9
Navsari	12.3	Surat	7.7
Vadodara	10.2	Banaskantha	7.4
Kachchh	10.2	Porbandar	7.0
Surendranagar	9.7	Jamnagar	6.7
Valsad	9.7	Anand	6.1
Junagadh	9.6	Kheda	6.0
Sabarkantha	9.6	Ahmedabad	6.0
Narmada	9.3	Rajkot	4.8
Amreli	8.5	Bhavnagar	4.3
Gandhinagar	8.4		

Source: Derived from a model of vulnerability, 2001.

The above analysis clearly indicates that the pattern of vulnerability can be explained by four major sets of determinants of vulnerability. Therefore, in order to reduce the vulnerability, it is important to address these drivers of vulnerability. Addressing these drivers can essentially reduce the risk due to climatic disaster by reducing vulnerability. The vulnerability index derived from these drivers represents relative position of districts within the state. The vulnerability index, in this context, helps in prioritising the district for development planning. Table 4.3 presents the vulnerability index.

As far as vulnerability pattern in Gujarat is concerned, the backward districts, namely Panchmahal, Dangs and Dohad, are at higher risk. Given a similar climatic hazard scenario within the state, these districts have higher chances of potential losses at the time of climatic extremities.

Applying Vulnerability Analysis to Calculate Risk

Let us examine the applicability of the vulnerability index for assessing the risk due to three selected disasters in Gujarat separately. The separate exercise was carried out because the mitigation strategies for each disaster are developed by separate departments. However, the method for deriving risk index has remained similar for all the three disasters.

Slow Onset Disasters: Drought

Drought is defined meteorologically as the situation in which rainfall in a particular year over a region is less than 75 per cent of long-term normal or average (Ray, 2000). Indian Meteorological Department (IMD) further divides the drought as per severity. If the deficit experienced in rainfall is in the range of 26 to 50 per cent , it is declared as moderate drought, while if the deficit is more than 50 per cent, severe drought is declared. Nonetheless, drought becomes visible when it is declared as agricultural and hydrological drought. Planning commission considers drought as a major factor contributing to regional imbalances in the development of the country. Therefore, it focuses on identifying the drought-prone areas for planning and implementation of various block development programmes. Such efforts can help in developing drought-prone areas in such a way that the adverse impact of drought can be minimised.

Table 4.4 indicates risk calculated from drought-prone area as well as vulnerability index. The drought-prone area index is derived from the

Department of Revenue, Government of Gujarat (GoG). It is important to mention here that the drought is declared by the state government in their relief manual. Such declaration helps in identifying the areas for implementing block area development plan. The declaration of drought-prone areas considers various factors into account. Such factors include the deficiency in rainfall, availability of ground and surface water, loss of agricultural production in the previous *kharif* and *rabi* season, the estimation of current year production, etc. If vulnerability index is incorporated into drought-prone area index, it provides an overall risk index due to drought. Now let us examine the importance and need of risk calculation.

Table 4.4

District-wise Variation in Risk due to Drought

No.	*District*	*Drought-Prone Area Index*	*Risk Index Due to Drought*	*Ranking as per Risk*
1	Panchmahals	10	133	1
2	Kachchh	10	102	2
3	Surendranagar	10	97	3
4	Amreli	11	93.5	4
5	Dohad	7	86.1	5
6	Jamnagar	9	60.3	6
7	Junagadh	6	57.6	7
8	Rajkot	12	57.6	8
9	Banaskantha	7	51.8	9
10	Vadodara	5	51	10
11	Narmada	4	37.2	11
12	Ahmedabad	6	36	12
13	Bharuch	4	32.8	13
14	Patan	4	31.6	14
15	Valsad	3	29.1	15
16	Bhavnagar	6	25.8	16
17	Porbandar	2	14	17
18	Dangs	1	13.2	18
19	Navsari	1	12.3	19
20	Sabarkantha	1	9.6	20
21	Gandhinagar	NA	-	-
22	Mehsana	NA	-	-
23	Surat	NA	-	-
24	Anand	NA	-	-
25	Kheda	NA	-	-

Source: Relief Department, Government of Gujarat, Risk is calculated by the author.

The calculation of risk is important in two ways. Firstly, it is important to prioritise the area of intervention for reducing regional imbalances at the time during drought. When vulnerability is considered into the risk calculation, the prioritisation of an area intervention also changes. For instance, let us compare the case of Panchmahal and Rajkot. Drought-prone area index shows that Rajkot district is more prone to drought compared to Panchmahal. However, risk index represents that Panchmahal is more vulnerable to drought than Rajkot. Therefore, if drought mitigation strategies need to be effective, they need to consider the regional imbalance into deliberation. An effective drought mitigation strategy needs to prioritise Panchmahal for implementing area development plan.

Secondly, the exercise of risk calculation is helpful in selecting the district for prioritising implementation when similar drought-prone areas are delineated in more than one district. For instance, Panchmahal, Kachchh and Surendranagar being similarly affected by drought, the vulnerability index helps in prioritising Panchmahal district for mitigating drought risk reduction strategies. Such prioritisation process helps in implementing the development programmes adeptly and efficiently. This is more useful in terms of cost effective investment at the time of constraint fund. Such efforts of calculating risk can help in the process of convergence of disaster risk reduction strategies with development planning strategies, as it also helps in reducing the regional disparity in the state.

Rapid Onset Disasters: Cyclone and Flood

The similar exercise was carried out for rapid onset disasters. Table 4.5 represents the risk calculation for flood. Flood being a rapid onset disaster needs a special attention during the process of flood management planning. Flood is experienced when the water overflows its natural or artificial banks onto dry lands. Although the flood can be natural or man made, the basic inflow of water has linkage with heavy rainfall. Therefore, it is normally considered as climate-induced disasters. The GoG has formed the Flood Control Cell to deal with flood situation. The Flood Control Cell has prepared a flood memorandum for managing the adverse effects of flood.

Table 4.5

District-wise Risk Due to Flood in Gujarat

No.	*District*	*Flood-Prone Area*	*Risk Index*	*Ranking as per Risk*
1	Dangs	100	1320	1
2	Navsari	100	1230	2
3	Valsad	100	970	3
4	Panchmahals	73	967	4
5	Junagadh	86	823	5
6	Kachchh	80	816	6
7	Surendranagar	80	776	7
8	Dohad	57	703	8
9	Porbandar	100	700	9
10	Vadodara	67	680	10
11	Gandhinagar	75	630	11
12	Amreli	73	618	12
13	Jamnagar	90	603	13
14	Sabarkantha	62	591	14
15	Bharuch	63	513	15
16	Kheda	80	480	16
17	Rajkot	100	480	17
18	Narmada	50	465	18
19	Patan	50	395	19
20	Bhavnagar	91	391	20
21	Ahmedabad	64	382	21
22	Surat	47	359	22
23	Anand	50	305	23
24	Banaskantha	33	247	24
25	Mehsana	NA	-	25

Source: Flood-prone area is derived from Flood Control Cell and Risk Index is calculated by the author.

The flood memorandum has been prepared basically for flood forecasting and preparedness for reducing the risk due to flood. Under the memorandum, the list of flood affected villages is prepared on the basis of discharge and gauge information collected from various sites of reservoirs and dams. It is assumed here that area under such villages is highly prone to flood. If the vulnerability index is merged into area prone index, one can derive the risk index due to flood. Although the flood relief and reconstruction depend on ground truth data at a time of flood, the disaster risk reduction strategies need a vulnerability analysis to be incorporated in risk calculation.

Table 4.6

District-wise Risk Due to Cyclone in Gujarat

No.	District	Cyclone-Prone Area	Risk Index	Ranking as per Risk
1	Panchmahals	100	1330	1
2	Dangs	100	1320	2
3	Vadodara	100	1020	3
4	Valsad	100	970	4
5	Narmada	100	930	5
6	Gandhinagar	100	840	6
7	Bharuch	100	820	7
8	Surat	100	770	8
9	Banaskantha	100	740	9
10	Junagadh	74.5	715.2	10
11	Jamnagar	91	609.7	11
12	Kheda	100	600	12
13	Porbandar	74.5	521.5	13
14	Mehsana	59.7	495.51	14
15	Patan	59.7	471.63	15
16	Kachchh	36.3	370.26	16
17	Amreli	32	272	17
18	Bhavnagar	62.6	269.18	18
19	Ahmedabad	25	150	19
20	Rajkot	21.3	102.24	20
21	Dohad	NA	-	-
22	Navsari	NA	-	-
23	Surendranagar	NA	-	-
24	Sabarkantha	NA	-	-
25	Anand	NA	-	-

Source: *Vulnerability Atlas*, Building Materials and Technology Promotion Council (BMTPC); Risk is calculated by the author.

Data on cyclone-prone area have been collected from the vulnerability atlas prepared by the BMTPC. The BMTPC classifies the cyclone-affected areas into various damage risk zone as per the speed of wind. Nonetheless, IMD considers that the cyclones which are having more than 60 km per hour wind speed are considered as severe cyclone and they cause damage in an affected area. As far as Gujarat state is concerned, the velocity of wind experienced in the past cyclones has remained higher than 158.4 km per hour. The risk of damage due to cyclone higher than 158.4 km per hour is considered to be very high and intensive. Therefore, the area prone to cyclones in Gujarat has considered only the frequency of occurrence of cyclones and has not accounted the intensity of cyclone. The risk calculation for cyclone is highly useful because many districts of Gujarat fall in a similar category of cyclone-prone areas. It is clear from Table 4.6 that the vulnerability differentiates the risk due to cyclone in districts facing similar damage due to disasters. The risk calculation is more important when planning for long-term mitigation strategies such as land use planning, improving livelihood of vulnerable people, resource planning, etc. in such a way that minimises the risk.

Composite Risk Index for Climate-Induced Disasters

This section attempts to derive a composite risk index for climate-induced disasters. It considers all the three climatic disasters, namely drought, flood and cyclone.

As discussed earlier, risk is a function of hazard and vulnerability index.

Therefore,

$$RI = fHi^{*}VI \qquad \ldots 1$$

where Hi is an index of hazard severity

VI is an index of vulnerability

$$Hi = \propto \underset{I=1}{\overset{3}{Hi}}$$

$$= (SQRT(\propto 1h1) * SQRT(\propto 2h2) * SQRT(\propto 3h3)) \qquad \ldots 2$$

where $\propto$ is a coefficient of an impact for particular hazard. Putting the value of Equation 2 in 1, we get

$$RI = (H1^{*}H2^{*}H3^{*}Vi)^{\wedge} 1/4.$$

Index is calculated for each district. The districts are ranked as per the risk index. Districts are, further, classified into groups facing varied degree of risk.

Table 4.7

Relative Composite Risk Index for Districts in Gujarat

No.	*District*	*Composite Risk Index*	*No.*	*District*	*Composite Risk Index*
1	Anand	3	14	Patan	12
2	Sabarkantha	5	15	Banaskantha	13
3	Mehsana	5	16	Bharuch	13
4	Navsari	5	17	Narmada	13
5	Dohad	6	18	Dangs	13
6	Surendranagar	6	19	Amreli	14
7	Surat	9	20	Valsad	14
8	Kheda	9	21	Kachchh	14
9	Gandhinagar	10	22	Jamnagar	14
10	Ahmedabad	11	23	Vadodara	14
11	Rajkot	11	24	Junagadh	14
12	Bhavnagar	11	25	Panchmahals	17
13	Porbandar	12			

Source: Calculated by the author.

The composite risk index measures the risk quantitatively. It provides direct comparison of relative overall risk of various districts in the state. This inclusive risk index underscores the fact that hazards can occur anywhere, and other characteristics of the region turn that hazard into disaster. The districts are further classified in four groups according to degree of vulnerability. Classification of districts considered the geometric mean and SD for dividing the districts and various groups. Mean plus and minus SD provided two cut-off points from which classes were constructed. The first group having the value higher than mean plus standard deviation was derived as highly vulnerable group. The second group was classified having risk index higher than mean but lower than mean plus standard deviation. About nine districts fall under this category. The third category accommodated four districts having risk index more than mean minus standard deviation but less than mean. The fourth category contains five districts which are non-vulnerable as the risk index is less than mean minus standard deviation.

Table 4.8

Classification of Districts as per Degree of Vulnerability Towards Climate-Induced Disasters

No.	*Highly Vulnerable >13.6*	*Moderately Vulnerable 10.8-13.6*	*Less Vulnerable 6.3-10.8*	*Not Vulnerable <6.3*
1.	Amreli	Ahmedabad	Surendranagar	Anand
2.	Valsad	Rajkot	Surat	Sabarkantha
3.	Kachchh	Bhavnagar	Kheda	Mehsana
4.	Jamnagar	Porbandar	Gandhinagar	Navsari
5.	Vadodara	Patan		Dohad
6.	Junagadh	Banaskantha		
7.	Panchmahals	Bharuch		
8.		Narmada		
9.		Dangs		

Source: Calculated by the author.

A relative index of risk calculation for all the districts indicates that 7 districts among 25 pose high risk towards climatic variability. Among them, 6 are located on the coastal region. Paradoxical to this, 6 districts do not face risk due to climatic disasters. These are mainly located in middle part of Gujarat.

In short, risk calculation due to climate-induced disasters has to focus on those regions which are currently experiencing development below some accepted norm. Currently less developed and vulnerable regions cope with higher risk while experiencing climate-induced disasters. Recalculating composite risk index periodically can help in assessing the trend of risk over a period of time. Such extended risk assessment can be useful in mainstreaming disaster risk reduction into development plans and strategies.

Conclusion

At global level, there is glut of literature showing that the seriousness and severity of adverse impact of disasters is increasing day by day. It is also evidently proved that the adverse effects are not proportionally experienced. The poor and vulnerable people are more at risk. The disaster scenario in Gujarat is also very much similar to that of global and national trend. In fact in spite of having good economic status in India, Gujarat is highly vulnerable to disaster risk because of variation in prevailing vulnerability within the state. Further, the economic development of the state make it more exposed towards disasters.

It is possible to assess the risk due to climatic variability. Such an exercise is quite useful at a state as well as district level planning. The development activities at district level have to incorporate the vulnerability analysis for assessing risk due to climatic disasters. This can ultimately help in formulating climate change adaptation strategies. It is very important to see that the development activities do not increase the vulnerability towards climatic variability. The district level planning can further look into the block level planning for further prioritisation of development activities. This will be an effective way to mainstream the disaster risk reduction and climate change adaptation strategies into the development goals.

5 Vulnerability Analysis at Micro Level

The study undertook several case studies to assess vulnerability at micro level. The vulnerability analysis at the micro level was carried out to serve two purposes. The first one is to ensure that whether the macro level vulnerability drivers are valid at micro level. The second purpose is to understand the vulnerability pattern existing at village level and addressing such vulnerability through development planning. The case studies have examined that how poor socioeconomic status and environmental degradation make the region more vulnerable towards climate-induced disasters. In this way, these case studies helped in understanding the relationship of climatic variability risk and vulnerability of the region.

The first step in this approach was to identify the representative sample. Ten villages/towns were selected for micro level study. In order to select representative sample, a stratified random sampling was adopted for the study. The strata were developed according to two major parameters. The first one is the type of climate-induced hazard and second one is level of vulnerability. Districts were classified according to vulnerability index derived at the state level. Table 5.1 shows the classes of existing vulnerability for all the districts.

Table 5.1

Selection of Case Studies

Vulnerability Classes		*Sample*		
		Urban	*Rural*	*Total*
10-14	6	1	1	2
7.1-10	13	1	3	5
4-7	6	1	1	2
	25			

Table 5.2

Selected District as per Hazards

	Cyclone	*Flood*	*Drought*
Urban	Jamnagar	Surat	Kutchh
Rural	Porbandar	Kheda	Surendranagar
Rural	Kutchchh	Porbandar	Sabarkantha

In this way, the districts were classified as per degree of vulnerability. The three classes were derived to obtain normal distribution pattern. Two districts were selected from the first and the third category, while five districts were selected from the second category. Such a selection includes 36 per cent of total districts in a sample.

The second step was the selection of villages and town (or city). It was necessary to select the case study facing various climate-induced disaster. Further both urban and rural case studies were selected from the above-selected districts. Out of nine selected case studies, three were urban regions and six were rural. The types of climates-induced hazards were also taken into consideration while selecting these case studies. One urban and two rural case studies were selected in such a way that each group of these three case studies face each of the climate-induced disasters cyclone and flood. While for drought, only two rural case studies were selected.

Selection of Households

The sample households (HHs) were selected strategically to fulfill the objectives of the study. We focused on vulnerable pocket for assessing and verifying vulnerability existing at micro level. Identification of such vulnerable pocket in the selected villages and urban slums was carried out through social mapping. Social mapping is an effective participatory rural appraisal (PRA) technique for sampling target households. Social mapping takes an account of visual representation of an inhabited area containing pertinent social information. It is carried out with the help of local resident's direct participation. Mapping of different areas with physical, socioeconomic and environmental parameters classified the habitation in various pockets. In fact, mapping generated a lot of zest among local people and it also worked as a good icebreaker. It was local community who identified the area (Faliya) which is highly vulnerable to climate-induced disasters. We selected few HHs to identify factors constructing vulnerability of these areas towards climate-induced disasters. This activity made the process of identification of HHs easy.

Table 5.3

Sample Households

Type of Disaster	*District*	*Taluka*	*Village/ Urban Slum*	*Rural/ Urban*	*Sample HH*
Drought	Sabarkantha	Megharaj	Vauna	R	15
	Surendranagar	Sayla	Dhamala	R	15
Flood	Kheda	Matar	Vansar	R	30
	Porbandar	Porbandar	Kesod lusala	R	15
	Surat	Surat	Limayat	U	15
Cyclone	Kuchh	Anjar	Bhadreshwar	R	15
	Porbandar	Porbandar	Miyani	R	15
	Kuchh	Gandhidham	Kandla	U	20
	Jamnagar	Jamnagar	Jamnagar	U	15
			Total		155

We selected minimum 15 HHs from each vulnerable pocket identified through social mapping. Overall 155 HHs were interviewed with the help of a questionnaire. A detailed HHs questionnaire was prepared to assess socioeconomic condition, livelihood patterns, risk due to hazards and vulnerability. In addition to it, various PRA techniques were used to identify drivers of vulnerability at micro level. The check list was prepared before conducting the field work.

Socioeconomic Status of the Selected Households

Socioeconomic status of the sample HHs is very poor. This is obvious because we focused on vulnerable people. Approximately 98 per cent of sample HHs belonged to backward class. Among backward classes, 17 per cent belonged to schedule caste, 3 per cent belonged to schedule tribe and 79 per cent belonged to other backward class. Schedule caste was observed dominantly in Vansar village. It was also present in Vauna village and Kandla port. Schedule tribe was reported from Vauna and Vansar villages.

Roughly 43 per cent of sample HHs possessed official below-poverty cards. However, our calculation on income data shows that about 80 per cent of sample HHs were living below poverty line. Our calculation on poverty rates is based on planning commission's revised norms, i.e., ₹ 965 per capita per month for urban areas and ₹ 781 per capita per month for rural areas. This big gap is more significant in terms of urban areas. In urban areas, majority of slum dwellers are migrants and do not possess below poverty line (BPL) card. Further, we observed that fishermen communities in Bhadreshwar do not live within village administrative boundary. They have to live on coast almost through out the year. Hence, they did not have any formal identity with them.

So far as religion is concerned, Kandla port, Bhadreshwar and Limayat are dominated by Muslim population. A little portion of sample was Christian in Vansar. Other than these, all others selected sample HHs followed Hindu religion. Briefly, the sample represented all the three major religions.

Although representation of male and female was almost equal in an overall sample, there is a variation observed among case studies. Vauna in Sabarkantha is dominated by female population. This is mainly because male members in this tribal village have to migrate seasonally in search of employment. It is important to note here that Kandla port has reported higher proportion of male population. Trade and commerce activities are dominated by male labourers and most of these labourers have migrated from other states such as Bihar, Orissa, etc.

Literacy level for the selected HHs is extremely less. About 65 per cent people are illiterates. Roughly 25 per cent people have obtained primary education. Only 9 per cent people have reported to have education level higher than primary level. Low education level makes these people unable to access the information sent before, during and after the disasters happenings. Therefore, precaution and proactive strategies to deal with disasters are almost nil among these communities. It is important to note here that illiteracy is more prevalent in urban areas compared to that in rural areas.

Main occupations of the selected villages included cultivation, fishery, animal husbandry, agricultural labours and other labours. About one third of sample HHs possess land. Among land holders, 60 per cent of farmers are marginal farmers who own land less than 2.5 acres. Remaining are medium farmers. Sample HHs from drought-prone villages Vauna and Dhamala as well as flood-prone villages Vansar and Keshod Lusala have reported farming as one of the major occupations. About half of the farmers do not avail irrigation facility and are rainfed. Agriculture in Dhamala and Kshod Lusala is completely rainfed, while farmers from Vauna (Panchmahal) and Vansar (Porbandar) villages use wells and tube wells as major irrigation sources. Fishing is the second major occupation reported by 45 HHs. They are mainly located in Kandla, Bhadreshwar and Miyani villages. In urban areas, most of the slum dwellers are involved as labours in nearby marketing yard or industrial activities. Flood-prone village Vansar has reported two third of sample HHs engaged as agricultural labours. Animal husbandry is reported in Vansar as secondary occupation.

With respect to social security, only 4 per cent have insurance and credit facility to maintain their livelihood during the time of crisis. Roughly 26 per cent have bank accounts to save money which they can use at the time of disasters. However, this is only observed in case studies from Saurashtra. People from Keshod Lusala and Miyani in Porbandar have reported their bank account. This is mainly due to many non-governmental organisations (NGOs) working actively in this area. However, drought-prone villages have not reported any crop insurance or credit availability at the time of disasters happening.

Physical Exposure

As far as housing quality is concerned, around 63 per cent selected HHs live in *kutcha* house. Twenty-three per cent of households live in semi-*pacca* and only 14 per cent live in *pacca* house that too provided under Indira Awas Yojana or Sardar Awas Yojana. This shows that majority of these rural and urban poor do not have capability to build their own *pacca* house. Houses in which they live are highly vulnerable to flood and cyclone. In urban centres, slums are located in low lying areas and remain flooded during every monsoon. Such urban flooding causes unhygienic surrounding and becomes a root cause for widespread diseases.

Access to Social Infrastructure

An easy access to safe and clean drinking water reduces vulnerability of people during disasters. Roughly 30 per cent people have access to water tap.

In urban areas, reverse osmosis (RO) plant is the only source for drinking water. This shows that the substantial proportion of HHS have access to clean drinking water. However, Dhamala in Surendranagar and Keshod Lusala in Porbandar do not have any public sources of drinking water. People in these two villages use hand pump and private wells for drinking water. Dhamala which is a drought-prone village suffers from severe water shortages during summer season. This makes the situation worse during the time of drought. Drinking water remains available only for six months. For remaining six months, community faces a real crisis of drinking water. Lushala on the other hand has good ground water level. However, water existing in wells and tube wells is saline with extremely high TDS (total dissolved salts). Hence, availability of drinking water is scarce in this village.

Similarly, status of sanitation is also extremely poor. Approximately 59 per cent of HHs do not use toilets and defecate in open ground. This unhygienic practice can increase the risk of spreading epidemic after flood disaster. This is prevalent in Keshod Lusala which is a flood-prone village. The village is located in low-lying area and gets flooded after every monsoon. In spite of such problem 80 per cent of HHs defecate in open in this village. Further, water shortages force people to defecate in an open space. In Dhamala, even though toilets are available, people do not use them due to acute water shortage. Urban areas are no better than rural areas in terms of hygienic sanitation. Urban slum dwellers of Bombay Dava Bazaar colony (Jamnagar) use open defecation near railway lines. Similar is the case of Kandla port workers.

So far as solid waste collection and management are concerned, 78 per cent of HHs throw their disposal in open ground. The case of urban areas is highly filthy. There is no system of waste collection and management existing in urban areas. Such unhygienic surroundings compel people to become highly vulnerable towards the risk due to flood and cyclone. Further drainage facility was not available to any of our selected case study except Miyani. In fact, Miyani also had only open drainage facility. Such poor and detrimental surroundings make these regions more vulnerable towards climate-induced disasters.

Access to health facility is relatively better in terms of physical accessibility. Urban areas are better than rural areas in terms of health facility. However, tribal villages are lagging behind in terms of physical access to health centres. Poor in Vauna have to walk at least five kilometres to reach the primary health centres. It is worth noticing here that people from Vansar village in developed Kheda district have to walk about 1 to 3 kilometres to reach to the heath centre.

Although all the rural case studies reported having primary school within approachable reach, physical access to middle school is quite far. The only village which reported having a middle school in near by areas is Keshod Lusala. Urban areas are better off when compared to rural areas as far as physical accessibility of educational institutions is considered.

Access to transportation services is a major problem in case of Vauna, a tribal village, and Bhadreshwar, a coastal fishing village. Geographical isolation in these two villages is an important factor making them vulnerable towards any climate-induced disaster.

In short, degree of accessibility to social infrastructure can vary the degree of vulnerability of community towards climate-induced disasters.

Drivers of Vulnerability at Micro Level

We used participatory vulnerability tool for identifying determinants of vulnerability towards the selected climate-induced disasters. We also carried out focused group discussion to support the statements provided by local community. Some of the important drivers of vulnerability, identified by micro level studies, are described at this juncture.

Vulnerability towards Cyclones

The case studies very clearly reveal that the most important driver of vulnerability is vulnerable people. Vulnerable people include those who live below poverty line, who are from marginalised community such as fishermen or who are urban areas labourers from unorganised sector. Fishermen community has to remain near the coast and therefore they cannot remain present within a village administrative boundary. As a result, they are cut off from the mainstream development projects. They do not have their identity and therefore, they are marginalised. In urban areas also, people who work as contract labourers do not receive any opportunity to sustain their livelihoods. In short, it is easy to identify the most vulnerable people in rural and urban areas by assessing their livelihood support.

Vulnerable surroundings, in terms of lack of access to basic services such as water supply, drainage and shelters, also emerged out as important determinants of vulnerability in Gujarat towards cyclone. Further, depletion of resources such as fishing stock and areas was also observed in the case study as an important determinant of vulnerability. One of the important determinants became apparent from the case study is the failure of institutional support. Although the state government and other NGOs are working for the welfare of such marginalised people, the help does not reach to such vulnerable people.

It seems that the vulnerable people are not able to access the institutional support provided to them. This may be because they do not have proper identity. Most of them are seasonal migrants and have to migrate throughout the year in search of livelihood support. Such people are neither involved in pre-disaster preparedness programme nor given any help after disaster happens.

This makes them unable to absorb the shocks of climatic disturbances such as cyclone. Loss of livelihood is the single major threat reported from the case studies. This makes the local community more vulnerable during the time of disaster.

Vulnerability towards Flood

It is obvious that the most vulnerable areas towards flood are located in low-lying areas. However, it is the new development project which makes these areas highly vulnerable towards flood. The new development projects (such as dam, road and wall) have blocked the natural channels of drainage of water run-offs. Such blocked natural channels of drainage have aggravated the water logging situation in flood-affected areas. Communities have reported that earlier water logging due to flood used to remain for few hours. Now water logging is a daily phenomenon during the monsoon season. Haphazard planning does not consider the disaster risk while developing new projects. Therefore, it is important to mainstream disaster risk reduction strategies into development planning.

Our case studies also reported the fact that the communities residing in low-lying area are usually migrants, casual labourers and marginalised poor. Such marginalised communities are unable to absorb the shocks of flood. They experience substantial disruption in their livelihood sources. Such people are highly subjected to wide spread diseases like malaria, chikungunya and dengue. This again makes them more vulnerable as they are unable to work for many days. The costs of medical expenditure and loss of livelihood income are quite high for these vulnerable people. Some of them go back to poverty cycle due to major disaster.

Further, these vulnerable people live in a pathetic living condition what we call as vulnerable surroundings. The quality of life is awfully poor as they do not have access to even basic needs. Polluted drinking water sources, open defecation, overflowing of drainage and garbage collection are the usual phenomena of an unhygienic surrounding. The status of health particularly of children and older people is at risk not only during the flood but also for a long time after the disaster has struck them.

Accessibility to formal institutional support and welfare programme is observed to be negligent. This is because most of these communities are living illegally on space. The only hope for them is if they are considered for vote bank, their status gets an improvement during election time.

Another typical problem observed specifically for urban slum was that their areas mainly fall under urban agglomeration. Such urban agglomerations were not considered as part of urban town planning schemes. Therefore, they neither obtain help from urban bodies nor are benefited by rural welfare programme. Hence, people in such areas still live in kutcha mud houses. Further their assets are also affected by frequent water logging due to flood. In rural areas, the agricultural labour, small landholding and animal husbandry are the major occupations of such communities. The small piece of agricultural land and animals are the major assets for these communities. Such assets are highly prone to flood situation.

The government support in these areas remains limited to awareness generation only. We observed few slogans by Gujarat State Disaster Management Authority (GSDMA). However, the risk sharing and risk reduction strategies were not reported in any case studies. A small number of large farmers had crop insurance. Otherwise, the communities were not aware of such risk reduction strategies.

Vulnerability towards Drought

As far as drought is concerned, we limited our case studies to rural areas only. This is mainly because the state has not experienced a severe hydrological drought for the last decade. We assume that urban dwellers are in the better position compared to rural dwellers at the time of drought. This is because the rural livelihood in completely dependent on water availability for agricultural activities. Therefore, we selected two case studies from drought-prone areas.

Vulnerable resources (mainly including scarce and degraded water resources) make people more vulnerable towards an external threat such as drought. Cultivation, labour work and animal husbandry are the limited available livelihood options for the selected case studies. Moreover, agriculture is rainfed and highly prone to climatic variability. The lack of rural diversified economic activities along with lower agricultural productivity reduces the adaptive capability of local community towards drought. Shortage of drinking water and encroached common grass land limits the development of animal husbandry. People in these villages do not have any saving and credit facilities. They do not have storage of food which they can use during the time of disaster. Government relief programmes are the only available option for the community during the time of drought.

The degradation of forest area is also observed as a key vulnerability determinant in tribal villages. Tribals' livelihood is completely dependent on fuel and other minor products from forest. The deterioration in forest quality has an adverse effect on community livelihood particularly during the drought seasons.

In short, lack of employment opportunity has emerged out as one of the major indicators of vulnerability towards drought. Inability to access inputs for agricultural sector is also another important factor which contributes to vulnerability of the local community. Inadequate access to water, fodder bank, certified seeds, credit and saving make marginal and small farmers vulnerable towards drought.

In short, the potential risk due to disaster not only emerges from the hazard but also increases due to the prevalent existing pattern of vulnerability. Therefore, the assessment of vulnerability is an important exercise to reduce vulnerability towards climate-induced disasters.

6 Conclusion and a Way Ahead

Gujarat is likely to face a substantial risk due to climate-induced disasters. The risk further increases because of non-sustainable path of development. With high economic growth and large disparity in vulnerability pattern, the risk is likely to increase manifold. Reducing this risk needs an immediate attention of policy makers in Gujarat. Addressing such risk needs to follow a concrete strategy of risk reduction.

Although there is recognition in the state that vulnerability plays an important role in determining risk due to climate-induced disaster, there is no serious attempt to consider it while disaster risk reduction strategies are concerned. Hence, it is important to assess vulnerability comprehensively. Vulnerability assessment needs a systematic approach to understanding the dynamics of vulnerability.

The analysis on climate-induced disaster showed that these disasters pose a great threat for the state of Gujarat. The risk due to such disasters is substantially high. Disaster induces setbacks in the process of development and subsequently slows down the progress towards achieving development goals. If not tackled efficiently, this risk can affect the development of the state.

The link between disaster risk and vulnerability is clear. The risk is highly dependent on prevailing vulnerability in the state. Therefore, vulnerability measurement should be a part of risk assessment.

Risk due to disaster is unevenly distributed within a society. It is obvious that highly vulnerable people are worst affected. Poor and marginalised victims of disaster possess limited capability to cope up or mitigate the risk. Such limitation forces them in a situation of malnutrition and subsequently to chronic poverty. This can ultimately affect the poor children and women, as they are left with poorer health and drudgery. It is, therefore, important to incorporate disaster risk reduction into ongoing poverty reduction programmes. Poverty reduction programmes, such as Mahatma Gandhi National Rural Employment Guarantee Act (MNREGA), need to be implemented during and after the phase of disaster in order to reduce the vulnerability towards climate-induced disasters.

Disasters destruct the infrastructure and adversely affect the growth and productivity. Access to poor social infrastructure plays a major role in increasing risk due to climate-induced disaster. Therefore, it is necessary that district health plans consider disaster risk reduction into the process of implementing health programmes. Allocation in health infrastructure can be made efficient when risk calculations are incorporated into the process of prioritising implementation. Disaster resistant infrastructure can reduce the risk effectively.

Economic exposure poses a great threat for disaster risk. Gujarat has remained at the forefront in terms of economic growth. The longest coastline offers a good locational advantage for industrial development. Further, economic incentives and liberal rules for establishing industry have made the state a prioritised destination for investment. Such processes enhance economic exposure in the state. It is, therefore, important to see that such processes do not increase the risk due to climate-induced disasters. A long-term land use planning and disaster resilient structures can reduce the risk due to climate-induced disasters.

Degradation of natural resources plays an important role in limiting livelihood options for poor during the time of disaster. Therefore, development of natural resources can have significant contribution in mitigating disaster risk in Gujarat. Risk reduction strategies can accelerate the process of regenerating environmental resources of the state. Such efforts can help in a long run as adaptation to climate change risk. Environment protection and natural resource development programmes should access the disaster risk reduction and adaptation fund for reducing vulnerability towards disaster in a short term and subsequently towards climate change threats.

Above efforts can help in mainstreaming disaster risk reduction into development programmes. Disaster management act very clearly indicates that all the line departments need to help Gujarat State Disaster Management Authority (GSDMA) during the time of disaster. However, it is also important that all these line departments facilitate disaster risk reduction activities during non-disaster time. In fact, each department can have their special intact fund for risk reduction strategies. This can help reducing vulnerability and subsequently help in trimming down the risk due to climate-induced disasters.

References

Adger, W.N. (1999). "Social Vulnerability to Climate Change and Extremes in Coastal Vietnam", *World Development* 27: 249-69.

AIDMI (2006). *Promoting Risk Reduction and Human Security Across India and South Asia*. Ahmedabad: All India Disaster Mitigation Institute.

Alagh, Y.K (1989). *Agro Climatic Planning: An Overview*. New Delhi: Planning Commission, Government of India.

———. (1991a). *Indian Development Planning and Policy*. Helsinki: WIDER Studies in Development Economics and Delhi: Vikas.

———. (1991b). Sustainable Development: From Concept to Action—Techniques for Planners. UNCED, RIO Conference.

———. (2006). "India 2020", in Journal of Quantitative Economics (New Series) 4(1): 1-14, January.

———. (2011). The Future of Indian Agriculture. NBT forthcoming.

Archaeological Survey of India (1979). *Lothal: 1955-62*. Vol. 1. Calcutta: Naba Mudran.

———. (1985). *Lothal: 1955-62*. Vol. 2. Calcutta: Naba Mudran.

Arya, Anand (2000). "Recent Development Towards Earthquake Risk Reduction in India", *Current Science* 79: 1270-71.

Bayer, Joanne (2006). *Risk and Vulnerability Programme*. Austria: International Institute for Applied Systems Analysis (IIASA).

Birkmann, John *et al.* (2006). Measuring Vulnerability to Natural Hazards. New Delhi: TERI Press.

Blaikie, P. *et al.* (1996). *At Risk: Natural Hazards, Peoples Vulnerability and Disasters*. London: Routledge.

Bogardi, J. and J. Birkmann (2004). "Vulnerability Assessment: The First Step Towards Sustainable Risk Reduction", in D. Malzahn and T. Plapp (eds.), *Disaster and Society—From Hazard Assessment to Risk Reduction*. Berlin: Logos Verlag. pp.75-82.

Bohle, H.G., T.E. Downing and M.J. Watts (1994). "Climate Change and Social Vulnerability: The Sociology and Geography of Food Insecurity", *Global Environmental Change* 4: 37-48.

Bollin, C., C. Cardenas, H. Hahn and K.S. Vatsa (2003). *Natural Disaster Network: Disaster Risk Management by Communities and Local Governments*. Washington, D.C.: Inter-American Development Bank.

Broadus, J.M. (2001). "Sea Level Rise and the Bangladesh and Nile Deltas", in J.X. Kasperson and R.E. Kasperson (ed.), *Global Environment Risk*. New York: UNU Press. pp.353-72.

Cannon, T., J. Twigg *et al.* (2005). *Social Vulnerability, Sustainable Livelihoods and Disasters*. Report to DFID Conflict and Humanitarian Assistance Department (CHAD) and Sustainable Livelihoods Support Office. London, DFID. p.63

Cardona, O.D. (1999). "Environmental Management and Disaster Prevention: Two Related Topics: A Holistic Risk Assessment and Management Approach", in J. Ingleton (ed.), *Natural Disaster Management*. London: Tudor Rose.

———. (2004). "The Need for Rethinking the Concepts of Vulnerability and Risk from a Holistic Perspective: A Necessary Review and Criticism for Effective Risk Management", in G. Bankoff, G. Frerks and D. Hilhorst (eds.), *Mapping Vulnerability: Disasters, Development and People*. London: Earthscan Publishers.

Chambers, R. and G. Conway (1992). "Sustainable Rural Livelihoods: Practical Concepts for the 21st Century", *IDS Discussion Paper* No. 296. Brighton: Institute of Development Studies.

Cutter, S.L., L. Barnes, M. Berry, C. Burton, E. Evans, E. Tate and J. Webb (2008). "A Place-Based Model for Understanding Community Resilience to Natural Disasters", *Global Environmental Change* 18(4): 598-606.

Davidson, R. (1997). *An Urban Earthquake Disaster Risk Index*. Report No. 121. The John A. Blume Earthquake Engineering Centre, Department of Civil Engineering, Stanford University.

Dilley, M. *et al.* (2005). *Natural Disaster Hotspot: A Global Risk Analysis*. New York: World Bank and Columbia University.

EPW (2003). *Domestic Product of State of India 1960-61 to 2000-01*. Mumbai: EPW Research Foundation.

FWWB (2002). *Social Security Program of People's Solidarity Association (PSA)*. India: Friends of Women's World Banking

GoG (2009). *Disaster Management Plan 2009-10: Flood Memorandum 2009*. Gandhinagar: Flood Control Cell.

———. (2005). Proceedings of the One-day Workshop on *Housing Insurance Awareness*. Conference at Imperial Palace Hotel, Rajkot. Gujarat State Disaster Management Authority (GSDMA).

———. (2003-04). *Gujarat State Insurance Annul Report*. Gujarat: Insurance Commissioner Office.

———. (2003). *Gujarat State Disaster Management Act, 2003 (No. 20)*. GSDMA.

———. (2003). *Annual Report 2003-04*. Ahmedabad: Gujarat State Disaster Management Authority.

———. (2001). *Earthquake Relief Operation in Gujarat.* Revenue Department.

———. (1982). *The Gujarat Relief Manual.* Revenue Department.

GoI (2006). *Vulnerability Atlas of India.* New Delhi: Building Materials and Technology Promotion Council, Ministry of Housing & Urban Poverty Alleviation.

———. (2006). *Towards Faster and More Inclusive Growth: An Approach to the 11th Five Year Plan.* New Delhi: Planning Commission.

———. (2005). *The Gujarat Vulnerability Atlas.* New Delhi: Building Materials & Technology Promotion Council.

———. (2004). *Calamity Relief in Twelfth Finance Commission Report* (Chapter-9).

———. (2001). *District Census Handbook.* New Delhi: Census of India.

———. (1989). *Agro Climatic Planning: An Overview.* New Delhi: Planning Commission (Authorship: Y.K. Alagh *et al.*).

———. (1981). *Development of Drought Prone Areas.* New Delhi: National Committee on the Development of Backward Areas, Planning Commission.

Gujarat Ecology Commission (2012). *Coastline of Gujarat: A Study on Ecology and Development: Potential of Beaches.* Gandhinagar.

Gupta, Navita (1999). *A Disaster Mitigation Policy Based on People's Response in the Flood Affected Areas: Case Study—Gurdaspur District, Punjab.* School of Planning, Centre for Environment Planning and Technology.

Gupta, P.K., S. Panigrahy and J.S. Parihar (2009). "Impact of Climate Change on the Major River Basins of India", in S. Panigrahi, S.S. Ray and J.S. Parihar (ed.), *Impact of Climate Change on Agriculture.* Ahmedabad: ISRO. pp.169-73.

Gupta, S.P. (1969). *Statistical Methods.* New Delhi: Sultan Chand & Sons.

Hirway, Indira and Darshini Mahadevia (2004). *Gujarat Human Development Report.* Ahmedabad: Mahatma Gandhi Labour Institute.

INCAA (2007). *India: Greenhouse Gas Emissions 2007.* New Delhi: Indian Network of Climate Change Assessment (INCAA), Ministry of Environment and Forest, Government of India.

———. (2010). *Climate Change and India: A 4x4 Assessment: A Sectoral and Regional Analysis for 2030s.* New Delhi: Indian Network of Climate Change Assessment (INCAA), Ministry of Environment and Forest, Government of India.

IPCC (2007). *Climate Change 2007: Impacts, Adaptation and Vulnerability: Summary for Policymakers.* Paris: Intergovernmental Panel on Climate Change.

Kapur, Anu (2010). *Vulnerable India: A Geographical Study of Disasters.* New Delhi: SAGE Publication.

Langridge, Ruth, Juliet Christian Smith and Kathleen A. Lohse (2006). "Access and Resilience: Analyzing the Construction of Social Resilience to the Threat of Water Scarcity", *Ecology & Society 11(2): 18.*

Lester, R.R. (2003). *Financing Rapid Onset Natural Disaster Losses in India*. Report No. 26844-I. World Bank

Lodhia, Shital (2010). *Sustainable Development of the Coastal Environment of Gujarat*. Germany: VDM Verlag Dr. Müller.

Malcom, Harper (2002). *Indian' Self- Help Groups and Bangladesh Grameen Bank Groups: A Comparative Analysis*. India: Friends of Women's World Banking.

Maskrey, Andrew (1989). *Disaster Mitigation: A Community Based Approach*. Oxford: Oxfam.

Miamidian, Eileen (2005) "Surviving Disasters and Supporting Recovery: A Guidebook for Microfinance Institutions", in *Disaster Risk Reduction Working Paper Series* No. 10. Washington, D.C.: World Bank

Mishra, A., Himani Upadhyay and G.J. Lingaraj (2011). "Climate Change and Agri-livelihoods", in B.K. Mohanty (ed.), *Economic Development in India: Issues and Challenges*. New Delhi: New Century Publications.

Nanavaty, Reema (1997). *How to Run a Viable Income Generating Activity*. Ahmedabad: Friends of Women's World Banking.

Nath, P.K. (2012). "Impact of Climate Change on Indian Economy", in S. Singh and S. Mohan Kumar (ed.), *Climate Change*. New Delhi: Rawat. pp.78-105.

O'Brien, K., R. Leichenko, V. Kelkar, H. Venema, G. Aandahl, H. Tompkins, A. Javed, S. Bhadwal, S. Barg, L. Nygaard, and J. West (2004). "Mapping Vulnerability to Multiple Stressors: Climate Change and Globalization in India", *Global Environmental Change* 14(4): 303-13.

Oliver-Smith, A. (2003). "Theorizing Vulnerability in a Globalized World: A Political Ecological Perspective", in G. Bankoff, G. Frerk and D. Hilhorst (eds.), *Mapping Vulnerability: Disasters, Development and People*. London: Earthscan. pp.10-24.

Parsuraman, S. and P.V. Unnikrishnan (2000). *India Disaster Report: Towards a Policy Initiative*. New Delhi: Oxford.

Philippe, Auffret (2003). "High Consumption Volatility: The Impact of Natural Disasters", *Policy Research Working Paper* 2962. Washington, D.C.: World Bank.

Ravindranath, N.H. *et al.* (2011). "Climate Change Vulnerability Profiles for North East India", *Current Science* 101(3): 284-394.

Ray, K., M. Mohanty and J.R. Chinchiolikar (2009). "Climate Variability over Gujarat, India", in S. Panigrahi, S.S. Ray and J.S. Parihar (ed.), *Impact of Climate Change on Agriculture*. Ahmedabad: ISRO. pp.38-43.

Ray, K.C.S. (2000). "Role of Drought Early Warning Systems for Sustainable Agricultural Research in India", in *Early Warning Systems for Drought Preparedness and Drought Management, Proceedings of an Expert Group Meeting*. Switzerland: World Meteorological Organization.

SAC (2010). *Desertification and Land Degradation: Atlas of India*. Ahmedabad: Space Applications Centre, Indian Space Research Organisation.

Singh, Katar and Ballabh Vishwa (2002). *Management of Disasters and Crisis Situations in India with Focus on the Poor*. Manila, Philippines: Asian Development Bank.

Suresh Kumar, S. (2012). "Climate Change: An Introspection into India's Official Data Base", in S. Singh and S. Mohan Kumar (ed.), *Climate Change*. New Delhi. pp.250-57

Turner, B.L. *et al.* (2003). "A Framework for Vulnerability Analysis in Sustainability Science", *Proceedings of the National Academy of Sciences* 100(14): 8074–79.

Twigg, John, (2006). *Hyogo and Other Indicator Frameworks: Convergence and Gaps*. November.

UNISDR (2005). "Hyogo Framework for Action 2005-2015: Building the Resilience of Nations and Communities to Disasters (HFA)", in *World Conference on Disaster Reduction*. January 8-22. Japan.

UNNATI (2004-05). *Annual Report*. Ahmedabad: UNNATI

UNDP (2011). *Human Development Report 2011—Sustainability and Equity: A Better Future for All*. New York: UNDP.

———. (2004). *Reducing Disaster Risk: A Challenge for Development*. New York: Bureau for Crisis Prevention and Recovery.

———. (2003). *Handbook for Estimating the Socioeconomic and the Environmental Effects of Disasters,* Volume 1. Mexico: Economic Commission for Latin America and the Caribbean.

———. (2000). *DRM Programme Information Booklet*. New York: UNDP.

United Nations (2005). *Hyogo Framework for Action 2005-2015: Building the Resilience of Nations and Communities to Disasters*. Geneva: UNISRD, A/CONF.206/6

UNISDR (2004). *Living with Risk: A Global Review of Disaster Reduction Initiatives*. Geneva: UN International Strategy for Disaster Reduction.

Villagran de Leon, J.C., (2006). "Vulnerability Assessment in the Context of Disaster-Risk, a Conceptual and Methodological Review", Source No. 4/20.

Wamsler, Christine (2006). "Operational Framework for Integrating Risk Reduction in Aid Organizations Working in Human Settlement Development", *Disaster Studies Working Paper* No. 14. London: Benfield Hazard Research Centre (BHRC).

Watts, M. and H.G. Bohle (1993). "The Space of Vulnerability: The Causal Structure of Hunger and Famine", *Progress in Human Geography* 17(1): 43-67.

Wisner, B., P. Blaikie, T. Cannon and I. Davis (2004). *At Risk: Natural Hazards, People's Vulnerability and Disasters*. Wiltshire: Routledge.

Websites:

http://www.fema.gov/

http://gujaratonline.com/newsroom/drought

www.gsdma.ac.in

http://www.gujaratstat.com/meteorologicaldata/22/naturalcalamities/179/stats.aspx

Annexures

Table 1

Distribution of Sample as per Backward Classes

Type of Disaster	*Name of Village*	*Rural/ Urban*	*Caste*				*Total Sample*
			SC	*ST*	*OBC*	*GEN*	*Total*
1	*2*	*3*	*4*	*5*	*6*	*7*	*8*
Drought	Vauna	R	6.67	20.00	66.67	6.67	15
	Dhamala	R	0.00	0.00	100.00	0.00	15
Flood	Vansar	R	70.00	3.33	20.00	6.67	30
	Kesod lusala	R	0.00	0.00	100.00	0.00	15
	Limayat	U	0.00	0.00	100.00	0.00	15
Cyclone	Bhadreshwar	R	0.00	0.00	100.00	0.00	15
	Miyani	R	0.00	0.00	100.00	0.00	15
	Kandla	U	20.00	0.00	80.00	0.00	20
	Jamnagar	U	0.00	0.00	100.00	0.00	15
	Total		16.77	2.58	78.71	1.94	155
			26	4	122	3	155

Table 2

Proportion of BPL Households

Type of	*Name of Village*	*Rural/Urban*	*BPL*		*Total*
			Yes	*No*	
1	*2*	*3*	*4*	*5*	*6*
Drought	Vauna	R	60.00	40.00	15
	Dhamala	R	40.00	60.00	15
Flood	Vansar	R	56.67	43.33	30
	Kesod lusala	R	66.67	33.33	15
	Limayat	U	46.67	53.33	15
Cyclone	Bhadreshwar	R	13.33	86.67	15
	Miyani	R	73.33	26.67	15
	Kandla	U	25.00	75.00	20
	Jamnagar	U	0.00	100.00	15
	Total		43.22	56.78	155
			67	88	

Table 3

Caste-wise Distribution of Sample Households

Type of Disaster	*Name of Village*	*Rural/ Urban*	*Religion*			*Total*
			Hindu	*Muslim*	*Christian*	
1	2	3	4	5	6	7
Drought	Vauna	R	100.00	0.00	0.00	15
	Dhamala	R	100.00	0.00	0.00	15
Flood	Vansar	R	96.67	0.00	3.33	30
	Kesod lusala	R	100.00	0.00	0.00	15
	Limayat	U	0.00	100.00	0.00	15
Cyclone	Bhadreshwar	R	0.00	100.00	0.00	15
	Miyani	R	100.00	0.00	0.00	15
	Kandla	U	20.00	80.00	0.00	20
	Jamnagar	U	100.00	0.00	0.00	15
			69.68	29.68	0.64	155
			108	46	1	155

Table 4

Distribution of Population

Type of Disaster	*Name of Village*	*Rural/Urban*	*Population*					
			Male	*Female*	*Children*	*Old Age*	*Disabled*	*Total*
1	2	3	4	5	6	7	8	9
Drought	Vauna	R	34.95	47.57	15.53	1.95	0	103
	Dhamala	R	40.30	37.31	16.42	5.97	0	67
Flood	Vansar	R	32.68	37.91	22.22	5.23	1.96	153
	Kesod lusala	R	43.33	33.33	21.67	1.67	0	60
	Limayat	U	43	41	31	0	0	115
			37.39	35.65	29.96	0	0	100
Cyclone	Bhadreshwar	R	39	53	21	11	2	126
			30.95	42.06	16.67	8.73	1.59	100
	Miyani	R	25	22	48	0	0	95
			26.32	23.16	50.52	0	0	100
	Kandla	U	50	36	53	0	0	139
			35.97	25.9	38.03	0	0	100
	Jamnagar	U	27	24	31	1	2	85
			31.76	28.24	36.47	1.18	2.35	100
	Total		323	328	258	27	7	943
			34.25	34.78	27.36	2.86	0.75	100

Table 5

Main Occupation of Sample Households

Type of Disaster	*Name of Village*	*Rural/ Urban*	*Main Occupation*				*Total*
			Agriculture	*Animal Hus.*	*Agriculture Lab.*	*Other Lab.*	
1	*2*	*3*	*4*	*5*	*6*	*7*	*8*
Drought	Vauna	R	15	0	0	0	15
			100	0	0	0	100
	Dhamala	R	12	0	0	3	15
			80	0	20	0	100
Flood	Vansar	R	6	2	19	3	30
			20	6.67	63.33	10	100
	Kesod lusala	R	12	0	0	3	15
			80	0	0	20	100
	Limayat	U	0	0	1	14	15
			0	0	6.67	93.33	100
Cyclone	Bhadreshwar	R	0	0	0	15	15
			0	0	0	100	100
	Miyani	R	0	0	0	15	15
			0	0	0	100	100
	Kandla	U	0	0	0	20	20
			0	0	0	100	100
	Jamnagar	U	0	0	0	15	15
			0	0	0	100	100
	Total		45	2	20	88	155
			29.03	1.29	12.9	56.78	100

Table 6

Type of Houses for Sample Households

Type of Disaster	*Name of Village*	*Rural/ Urban*	*House*			*Total*
			Kutcha	*Semi-Pacca*	*Pacca*	
1	*2*	*3*	*4*	*5*	*6*	*7*
Drought	Vauna	R	86.67	6.67	6.67	15
	Dhamala	R	20	53.33	26.67	15
Flood	Vansar	R	63.33	26.67	10	30
	Kesod lusala	R	46.67	53.33	0	15
	Limayat	U	100	0	0	15
Cyclone	Bhadreshwar	R	100	0	0	15
	Miyani	R	6.67	20	73.33	15
	Kandla	U	95	0	5	20
	Jamnagar	U	5	8	2	15
			33.33	53.33	13.34	100
	Total		97	36	22	155
			62.58	23.23	14.19	100

Table 7

Type of Employment

Type of Disaster	Name of Village	Rural/ Urban	No. of Employment				Total Employment
			Agriculture	Animal Hus.	Agriculture Lab.	Other Lab.	
1	2	3	4	5	6	7	8
Drought	Vauna	R	48	0	0	0	48
			100	0	0	0	100
	Dhamala	R	32	0	0	9	41
			78.05	0	0	21.95	100
Flood	Vansar	R	12	5	36	4	57
			21.05	8.77	63.16	7.01	100
	Kesod lusala	R	36	0	0	7	43
			83.73	0	0	16.27	100
	Limayat	U	0	0	2	32	34
			0	0	5.88	94.12	100
Cyclone	Bhadreshwar	R	0	0	0	55	55
			0	0	0	100	100
	Miyani	R	0	0	0	24	24
			0	0	0	100	100
	Kandla	U	0	0	0	45	45
			0	0	0	100	100
	Jamnagar	U	0	0	0	31	31
			0	0	0	100	100
	Total		128	5	38	207	378
			33.86	1.32	10.05	54.77	100

Table 8

Subsidiary Occupation among Sample Households

Type of Disaster	Name of Village	Rural/ Urban	Subsidiary Occupation				Total Employment
			Agriculture	Animal Hus.	Agriculture Lab.	Other Lab.	
1	2	3	4	5	6	7	8
Drought	Vauna	R	0	10	3	1	14
			0	71.43	21.43	7.14	100
	Dhamala	R	0	0	0	6	6
			0	0	0	100	100
Flood	Vansar	R	1	13	2	0	16
			6.25	81.25	12.5	0	100
	Kesod lusala	R	0	5	0	3	8
			0	62.5	0	37.5	100
	Limayat	U	0	0	0	0	0
			0	0	0	0	0
Cyclone	Bhadreshwar	R	0	0	0	0	0
			0	0	0	0	0
	Miyani	R	0	0	0	0	0
			0	0	0	0	0
	Kandla	U	0	0	0	0	0
			0	0	0	0	0
	Jamnagar	U	1	0	0	1	2
			50	0	0	50	100
	Total		2	28	5	11	46
			4.35	60.87	10.87	23.91	100

Table 9

Distribution of Sample Households as per Annual Income

Type of Disaster	Name of Village	Rural/Urban	Total Annual Income Household (₹ in 000)					
			0-25	25-50	50-75	75-1 lakh	1 lakh+	Total
1	2	3	4	5	6	7	8	9
Drought	Vauna	R	6	6	3	0	0	15
			40	40	20	0	0	100
	Dhamala	R	3	8	4	0	0	15
			20	53.33	26.67	0	0	100
Flood	Vansar	R	27	3	0	0	0	30
			90	10	0	0	0	100
	Kesod lusala	R	4	3	8	0	0	15
			26.67	20	53.33	0	0	100
	Limayat	U	1	4	1	9	0	15
			6.67	26.67	6.67	60	0	100
Cyclone	Bhadreshwar	R	4	2	4	3	2	15
			6.67	13.33	26.67	20	13.33	100
	Miyani	R	6	3	3	2	1	15
			40	20	20	13.33	6.67	100
	Kandla	U	7	3	6	4	0	20
			35	15	30	20	0	100
	Jamnagar	U	1	4	8	2	0	15
			6.67	26.67	53.33	13.33	0	100
	Total		59	36	37	20	3	155
			38.06	23.23	23.87	12.9	1.94	100

Table 10

Education Level among Sample Households

Type of Disaster	Name of Village	Rural/ Urban	Education					
			Illiterate	1-7	8-11	12	College	Total
1	2	3	4	5	6	7	8	9
Drought	Vauna	R	48	30	7	10	2	97
			49.48	30.93	7.22	10.31	2.06	100
	Dhamala	R	31	17	3	2	2	55
			56.36	30.91	5.45	3.64	3.64	100
Flood	Vansar	R	84	26	18	4	0	132
			63.64	19.7	13.64	3.02	0	100
	Kesod lusala	R	40	12	1	1	0	54
			74.08	22.22	1.85	1.85	0	100
	Limayat	U	69	29	3	0	0	101
			68.32	28.71	2.97	0	0	100
Cyclone	Bhadreshwar	R	63	31	11	2	0	107
			58.88	28.97	10.28	1.87	0	100
	Miyani	R	45	23	8	1	1	78
			57.69	29.49	10.26	1.28	1.28	100
	Kandla	U	101	28	1	1	0	131
			77.1	21.37	0.76	0.77	0	100
	Jamnagar	U	51	10	0	0	0	61
			83.61	16.39	0	0	0	100
	Total		532	206	52	21	5	816
			65.2	25.25	6.37	2.57	0.61	100

Table 11

Landholders among Sample Households

Type of Disaster	*Name of Village*	*Rural/ Urban*	Agr. Land (Vigha)			*Total*
			0-5	5-10	10+	
1	*2*	*3*	*4*	*5*	*6*	*7*
Drought	Vauna	R	10	2	0	12
			83.33	16.67	0	100
	Dhamala	R	2	2	6	10
			20	20	60	100
Flood	Vansar	R	19	0	0	19
			100	0	0	100
	Kesod lusala	R	1	6	5	12
			8.33	50	41.67	100
	Limayat	U	0	0	0	0
			0	0	0	0
Cyclone	Bhadreshwar	R	0	0	0	0
			0	0	0	0
	Miyani	R	0	0	0	0
			0	0	0	0
	Kandla	U	0	0	0	0
			0	0	0	0
	Jamnagar	U	0	0	0	0
			0	0	0	0
	Total		32	10	11	53
			60.38	18.87	20.75	100

Table 12

Irrigated Land among Sample Households

Type of Disaster	*Name of Village*	*Rural/ Urban*	*Irrigated Agr.Land (Vigha)*			*Total*
			0-5	*5-10*	*10+*	
1	*2*	*3*	*4*	*5*	*6*	*7*
Drought	Vauna	R	10	1	0	11
			90.91	9.09	0	100
	Dhamala	R	0	0	0	0
			0	0	0	0
Flood	Vansar	R	19	0	0	19
			100	0	0	100
	Kesod lusala	R	0	0	0	0
			0	0	0	0
	Limayat	U	0	0	0	0
			0	0	0	0
Cyclone	Bhadreshwar	R	0	0	0	0
			0	0	0	0
	Miyani	R	0	0	0	0
			0	0	0	0
	Kandla	U	0	0	0	0
			0	0	0	0
	Jamnagar	U	0	0	0	0
			0	0	0	0
	Total		29	1	0	30
			96.67	3.33	0	100

Table 13

Source of Irrigation

Type of Disaster	*Name of Village*	*Rural/ Urban*	*Irrigation Source*				*Total*
			Well	*Tube Well*	*Canal*	*Other (Rent)*	
1	*2*	*3*	*4*	*5*	*6*	*7*	*8*
Drought	Vauna	R	1	9	0	1	11
			9.09	81.82	0	9.09	100
	Dhamala	R	0	0	0	0	0
			0	0	0	0	10
Flood	Vansar	R	6	13	0	0	19
			31.58	68.42	0	0	100
	Kesod lusala	R	0	0	0	0	0
			0	0	0	0	0
	Limayat	U	0	0	0	0	0
			0	0	0	0	0
Cyclone	Bhadreshwar	R	0	0	0	0	0
			0	0	0	0	0
	Miyani	R	0	0	0	0	0
		0	0	0	0	0	
	Kandla	U	0	0	0	0	0
			0	0	0	0	0
	Jamnagar	U	0	0	0	0	0
			0	0	0	0	0
	Total		7	22	0	1	30
			23.33	73.33	0	3.34	

Table 14

Availability of Water for Irrigation

Type of Disaster	*Name of Village*	*Rural/ Urban*	*Available Water (month)*				*Total*
			0-3	*3-6*	*6-9*	*9-12*	
1	*2*	*3*	*4*	*5*	*6*	*7*	*8*
Drought	Vauna	R	1	2	2	6	11
			9.09	18.18	18.18	54.55	100
	Dhamala	R	0	0	0	0	0
			0	0	0	0	0
Flood	Vansar	R	0	0	0	19	19
			0	0	0	100	100
	Kesod lusala	R	0	0	0	0	0
			0	0	0	0	0
	Limayat	U	0	0	0	0	0
			0	0	0	0	0
Cyclone	Bhadreshwar	R	0	0	0	0	0
			0	0	0	0	0
	Miyani	R	0	0	0	0	0
			0	0	0	0	0
	Kandla	U	0	0	0	0	0
			0	0	0	0	0
	Jamnagar	U	0	0	0	0	0
			0	0	0	0	0
	Total		1	2	2	25	30
			3.33	6.67	6.67	83.33	100

Table 15

Source of Drinking Water for Selected Households

Type of Disaster	*Name of Village*	*Rural/ Urban*	*Drinking Water Source*					
			Tap	*Hand Pump*	*Well*	*Tube Well*	*Other (RO Plant)*	*Total*
1	*2*	*3*	*4*	*5*	*6*	*7*	*8*	*9*
Drought	Vauna	R	0	0	3	12	0	15
			0	0	20	80	0	100
	Dhamala	R	0	0	0	15	0	15
			0	0	0	100	0	100
Flood	Vansar	R	28	0	2	0	0	30
			93.33	0	6.67	0	0	100
	Kesod lusala	R	0	15	0	0	0	15
			0	100	0	0	0	100
	Limayat	U	15	0	0	0	0	15
			100	0	0	0	0	100
Cyclone	Bhadreshwar	R	0	0	0	0	15	15
			0	0	0	0	100	100
	Miyani	R	0	0	0	0	15	15
			0	0	0	0	100	100
	Kandla	U	0	0	0	0	20	20
		0	0	0	0	100	100	
	Jamnagar	U	2	6	5	0	2	15
			13.33	40	33.33	0	13.33	100
	Total		45	21	10	27	52	155
			29.03	13.55	6.45	17.42	33.55	100

Table 16

Availability of Drinking Water for Sample Households

Type of Disaster	*Name of Village*	*Rural/Urban*	*Water Available (month)*				*Total*
			0-3	*3-6*	*6-9*	*9-12*	
1	*2*	*3*	*4*	*5*	*6*	*7*	*8*
Drought	Vauna	R	0	0	0	15	15
			0	0	0	100	100
	Dhamala	R	0	15	0	0	15
			0	100	0	0	100
Flood	Vansar	R	0	0	0	30	30
			0	0	0	100	100
	Kesod lusala	R	0	0	0	15	15
			0	0	0	100	100
	Limayat	U	0	0	0	15	15
			0	0	0	100	100
Cyclone	Bhadreshwar	R	0	0	0	15	15
			0	0	0	100	100
	Miyani	R	0	0	0	15	15
			0	0	0	100	100
	Kandla	U	20	0	0	0	20
			100	0	0	0	100
	Jamnagar	U	0	0	0	15	15
			0	0	0	100	100
	Total		20	15	0	120	155
			12.9	9.68	0	77.42	100

Table 17

Type of Sanitation Facility Used by Sample Households

Type of Disaster	Name of Village	Rural/Urban	Sanitation			Total
			Individual	Common	Open Defecation	
1	2	3	4	5	6	7
Drought	Vauna	R	6	9	0	15
			40	60	0	100
	Dhamala	R	0	0	15	15
			0	0	100	100
Flood	Vansar	R	17	1	12	30
			56.67	3.33	40	100
	Kesod lusala	R	3	0	12	15
			20	00	80	100
	Limayat	U	0	15	0	15
			0	100	0	100
Cyclone	Bhadreshwar	R	0	0	15	15
			0	0	100	100
	Miyani	R	3	0	12	15
			20	0	80	100
	Kandla	U	0	10	10	20
			0	50	50	100
	Jamnagar	U	0	0	15	15
			0	0	100	100
	Total		29	35	91	155
			18.71	22.58	58.71	100

Table 18

Practices of Solid Waste Management in Selected Case Studies

Type of Disaster	Name of Village	Rural/ Urban	Solid Waste Management			Open Ground	Total
			Dustbin	Common Dustbin	Gram Panchayat or Municipal Workers		
1	2	3	4	5	6	7	8
Drought	Vauna	R	1	0	1	13	15
			6.67	0	6.67	86.66	100
	Dhamala	R	0	0	0	15	15
			0	0	0	100	100
Flood	Vansar	R	23	1	0	6	30
			76.67	3.33	0	20	100
	Kesod lusala	R	0	0	0	15	15
			0	0	0	100	100
	Limayat	U	0	0	0	15	15
			0	0	0	100	100
Cyclone	Bhadreshwar	R	0	0	0	15	15
			0	0	0	100	100
	Miyani	R	8	0	0	7	15
			53.33	0	0	46.67	100
	Kandla	U	0	0	0	20	20
			0	0	0	100	100
	Jamnagar	U	0	0	0	15	15
			0	0	0	100	100
	Total		32	1	1	121	155
			20.65	0.65	0.65	78.06	100

Table 19

Facility of Drainage among Sample Households

Type of Disaster	*Name of Village*	*Rural/Urban*	*Drainage*			*Total*
			Close	*Open*	*No Facility*	
1	*2*	*3*	*4*	*5*	*6*	*7*
Drought	Vauna	R	0	15	0	15
			0	100	0	100
	Dhamala	R	0	15	0	15
			0	100	0	100
Flood	Vansar	R	0	30	0	30
			0	100	0	100
	Kesod lusala	R	0	15	0	15
			0	100	0	100
	Limayat	U	0	15	0	15
			0	100	0	100
Cyclone	Bhadreshwar	R	0	15	0	15
			0	100	0	100
	Miyani	R	7	8	0	15
			46.67	53.33	0	100
	Kandla	U	0	20	0	20
			0	100	0	100
	Jamnagar	U	0	15	0	15
			0	100	0	100
	Total		7	148	0	155
			4.52	95.48	0	100

Table 20

Access to Social Security

Type of Disaster	*Name of Village*	*Rural/ Urban*	*Social Security*				*Total*
			Voting Card	*BPL Card*	*Bank Account*	*Insurance/ Credit facility*	
1	*2*	*3*	*4*	*5*	*6*	*7*	*8*
Drought	Vauna	R	15	15	0	0	15
			100	100	0	0	100
	Dhamala	R	15	0	2	1	15
			100	0	13.33	6.67	100
Flood	Vansar	R	30	30	3	1	30
			100	100	10	3.33	100
	Kesod lusala	R	15	15	8	0	15
			100	100	53.33	0	100
	Limayat	U	15	14	6	0	15
			100	93.33	40	0	100
Cyclone	Bhadreshwar	R	15	11	9	3	15
			100	73.33	60	20	100
	Miyani	R	15	10	4	0	15
			100	66.67	26.67	0	100
	Kandla	U	20	19	9	1	20
			100	95	45	0.5	100
	Jamnagar	U	2	15	0	0	15
			13.33	100	0	0	100
	Total		142	119	41	6	155
			91.61	76.77	26.45	3.87	100

Table 21

Access to Primary Health Centre

Type of Disaster	*Name of Village*	*Rural/Urban*	*Primary Health (km)*				*Total*
			<1	*1-3*	*3-5*	*>5*	
1	*2*	*3*	*4*	*5*	*6*	*7*	*8*
Drought	Vauna	R	0	0	0	15	15
			0	0	0	100	100
	Dhamala	R	15	0	0	0	15
			100	0	0	0	100
Flood	Vansar	R	0	30	0	0	30
			0	100	0	0	100
	Kesod lusala	R	15	0	0	0	15
			100	0	0	0	100
	Limayat	U	15	0	0	0	15
			100	0	0	0	100
Cyclone	Bhadreshwar	R	15	0	0	0	15
			100	0	0	0	100
	Miyani	R	15	0	0	0	15
			100	0	0	0	100
	Kandla	U	20	0	0	0	20
			100	0	0	0	100
	Jamnagar	U	0	15	0	0	15
			0	100	0	0	100
	Total		80	60	0	15	155
			51.61	38.71	0	9.67	100

Table 22

Access to Primary School

Type of Disaster	*Name of Village*	*Rural/Urban*	*Primary School (km)*				*Total*
			<1	*1-3*	*3-5*	*>5*	
1	*2*	*3*	*4*	*5*	*6*	*7*	*8*
Drought	Vauna	R	15	0	0	0	15
			100	0	0	0	100
	Dhamala	R	15	0	0	0	15
			100	0	0	0	100
Flood	Vansar	R	30	0	0	0	30
			100	0	0	0	100
	Kesod lusala	R	15	0	0	0	15
			100	0	0	0	100
	Limayat	U	0	15	0	0	15
			0	100	0	0	100
Cyclone	Bhadreshwar	R	15	0	0	0	15
			100	0	0	0	100
	Miyani	R	15	0	0	0	15
			100	0	0	0	100
	Kandla	U	0	20	0	0	20
			0	100	0	0	100
	Jamnagar	U	0	15	0	0	15
			0	100	0	0	100
	Total		105	50	0	0	155
			67.74	32.26	0	0	100

Table 23

Access to Middle School

Type of Disaster	*Name of Village*	*Rural/Urban*	*Middle School*				*Total*
			<1	*1-3*	*3-5*	*>5*	
1	*2*	*3*	*4*	*5*	*6*	*7*	*8*
Drought	Vauna	R	0	0	0	15	15
			0	0	0	100	100
	Dhamala	R	0	0	0	15	15
			0	0	0	100	100
Flood	Vansar	R	0	0	0	30	30
			0	0	0	100	100
	Kesod lusala	R	0	15	0	0	15
			0	100	0	0	100
	Limayat	U	15	0	0	0	15
			100	0	0	0	100
Cyclone	Bhadreshwar	R	0	0	0	15	15
			0	0	0	100	100
	Miyani	R	0	0	15	0	15
			0	0	100	0	100
	Kandla	U	0	0	0	20	20
			0	0	0	100	100
	Jamnagar	U	0	15	0	0	15
			0	100	0	0	100
	Total		15	30	15	95	155
			9.68	19.35	9.68	61.29	100

Table 24

Access to College

Type of Disaster	*Name of Village*	*Rural/Urban*	*College*				*Total*
			<1	*1-3*	*3-5*	*>5*	
1	*2*	*3*	*4*	*5*	*6*	*7*	*8*
Drought	Vauna	R	0	0	0	15	15
			0	0	0	100	100
	Dhamala	R	0	0	0	15	15
			0	0	0	100	100
Flood	Vansar	R	0	0	0	30	30
			0	0	0	100	100
	Kesod lusala	R	0	0	0	15	15
			0	0	0	100	100
	Limayat	U	0	0	15	0	15
			0	0	100	0	100
Cyclone	Bhadreshwar	R	0	0	0	15	15
			0	0	0	100	100
	Miyani	R	0	0	0	15	15
			0	0	0	100	100
	Kandla	U	0	0	0	15	20
			0	0	0	100	100
	Jamnagar	U	0	0	15	0	15
			0	0	100	0	100
	Total		0	0	30	125	155
			0	0	19.35	80.65	100

Table 25

Access to Bus Stop

Type of Disaster	*Name of Village*	*Rural/Urban*	*Bus Stop*				*Total*
			<1	*1-3*	*3-5*	*>5*	
1	*2*	*3*	*4*	*5*	*6*	*7*	*8*
Drought	Vauna	R	0	15	0	0	15
			0	100	0	0	100
	Dhamala	R	15	0	0	0	15
			100	0	0	0	100
Flood	Vansar	R	30	0	0	0	30
			100	0	0	0	100
	Kesod lusala	R	15	0	0	0	15
			100	0	0	0	100
	Limayat	U	15	0	0	0	15
			100	0	0	0	100
Cyclone	Bhadreshwar	R	0	15	0	0	15
			0	100	0	0	100
	Miyani	R	15	0	0	0	15
			100	0	0	0	100
	Kandla	U	20	0	0	0	20
			100	0	0	0	100
	Jamnagar	U	15	0	0	0	15
			100	0	0	0	100
	Total		125	30	0	0	155
			80.65	19.35	0	0	100